ERRATA

The publisher regrets that through an error in printing certain lines were omitted from the text and two paragraphs were repeated. Caption directions on page 115 are incorrect.

On page 108, the last line should continue onto page 110 as follows:
Clearly, the main **threat to orangutans consists in the destruction of their habitat. The illegal pet trade is an additional threat precisely because orangutans are so like humans, seeming so "cute" as infants and juveniles. They are especially prized for some reason in Taiwan, where, between 1988 and 1993, more orangutans were sold or traded** illegally than are housed in all the world's zoos.

On page 113, the last eight lines are repeated at the top of page 117.

On page 115, the first caption refers to the photograph directly above; the middle caption refers to the photograph at the left; and the bottom caption refers to the photograph at the top.

Also on page 117, at the bottom, the last line should continue onto page 120 as follows:
These metals then run off in the wash water and contaminate the tons of **silt that pour into the Sekonyer River. The once transparent black waters of the Sekonyer now look like café au lait. Now turbid and polluted, the Sekonyer serves both as the highway and bathtub for thousands of people living downstream, who depend on the river for drinking water and hygiene. What damage gold panning and surface mining will do to these people is as yet unknown, but the already obvious decline in bird life along the river's edge foretells a heavy price.**

Fortunately, no mining occurs directly upstream of Camp Leakey. The waters of the Sekonyer Kanan flow clear and black past the camp. Until the Sekonyer Kanan meets the main Sekonyer, the two streams of water are visibly distinct, but in a matter of a few yards the transparent waters of the tributary disappear into the cloudy murk of the polluted main river.

On page 123, the last thirteen lines are repeated at the top of page 124.

On page 130, in the top caption, the word "studies" should read "studied."

The authors are in no way responsible for these regrettable errors and the publisher apologizes for any inconvenience to the reader.

ORANGUTAN ODYSSEY

ORANGUTAN ODYSSEY

by Biruté M. F. Galdikas

and Nancy Briggs

Introduction by Jane Goodall

Photographs by Karl Ammann

HARRY N. ABRAMS, INC., PUBLISHERS

Dedicated to Louis Leakey
and all of the orangutans
who made this book possible.

EDITOR: Robert Morton
DESIGNER: Raymond P. Hooper

Library of Congress Cataloging-in-Publication Data

Galdikas, Biruté Marija Filomena.
Orangutan odyssey / Biruté M.F. Galdikas and Nancy Briggs ;
photographs by Karl Ammann ; introduction by Jane Goodall.
p. cm.
ISBN 0-8109-3694-1
1. Orangutan—Borneo. I. Briggs, Nancy (Nancy Erickson)
II. Title.
QL737.P96G356 1999
599.88'3'095983—dc21 99-25446

Published in 1999 by Harry N. Abrams, Incorporated, New York

Printed and bound in Hong Kong

Harry N. Abrams, Inc.
100 Fifth Avenue
New York, N.Y. 10011
www.abramsbooks.com

CONTENTS

INTRODUCTION

by Jane Goodall

For the past three decades Biruté Galdikas has been studying, conserving, and protecting orangutans and tropical rain forest in Kalimantan (Indonesian Borneo), an endeavor that has caught the attention of the whole world. In a series of articles and books, she has painstakingly documented orangutan behavior and ecology previously unknown to science. Her recent popular book *Reflections of Eden*, a fascinating account of her life with the orangutans, left me in awe of the important scientific work she has accomplished with the elusive and endangered orangutan. This book, Biruté's *Orangutan Odyssey*, expands the vision displayed in *Reflections of Eden*. Biruté deserves international praise and support for her dedication and devotion to saving these precious creatures, and for providing an invaluable scientific record that will probably never be repeated in the history of science.

Jane Goodall at the Gombe Stream Reserve in Africa. (Photograph by Ken Regan.)

Biruté, Dian Fossey, and I shared the same mentor, Louis Leakey (1903–1972), the great paleoanthropologist whose work revised our understanding of human evolution. I know Louis would be very proud of what Biruté has accomplished. I also know that he would be very proud of the fact that, against great odds, Biruté Galdikas continues to persevere in her struggle to save orangutans from extinction. It is not easy work. Biruté and I are sisters in the struggle to preserve great apes in the wild.

In 1991 I fulfilled a dream when I was able to visit Biruté's camp, Camp Leakey, in the heart of the Bornean rain forest and see with my own eyes what Biruté has accomplished. She had successfully habituated dozens of wild orangutans and was amassing long-term longitudinal records of their life histories based on meticulously detailed day-to-day observations by herself and her Indonesian staff. Her landmark discoveries concerning the long orangutan birth interval, orangutan ecology, and male-male competition have helped change our understanding of orangutans. She has, for example, documented that orangutan females in her area only give birth, on average, once every eight years and that orangutans eat more than four hundred different food types. Biruté also was conducting long-term phenological studies of trees in carefully censused botanical plots. Several Indonesian and European postgraduate students were observing proboscis monkeys as well as orangutans. It was only when I spoke with some young and energetic Indonesian ecologists and biologists, all former students of Biruté's who had gone on to get their Ph.D.s, that I finally fully realized how much influence Biruté was having on the young people and the future of her adopted homeland, Indonesia, and on people throughout the world.

As the millennium approaches, all great apes, whether in Africa or Asia, face an increasingly uncertain future. In Africa destruction of rain forest, poaching, and the rise of commercial "bush meat" trade constitute deadly large-scale threats against chimpanzees and gorillas. Orangutans also are caught in a political stranglehold, with palm oil plantations, mining, illegal deforestation, and greed destroying orangutan habitat. Biruté has always been in a battle with the powers of darkness. She is a thorn in the side of those who decimate Borneo's rain forest and that is why her struggles never seem to cease.

In Borneo and Sumatra, the only two places where wild orangutans are found, the conversion of tropical rain forests to plantations spells disaster for the arboreal orange apes who need primary forest to survive. In 1997 and 1998 the numerous fires that spread a heavy haze over Southeast Asia destroyed millions of hectares of forest, leaving countless orangutans homeless. Biruté and the Orangutan Foundation International (OFI), headquartered in Los Angeles, and active in four other countries, have organized the rescue of orphaned orangutans, have returned more than two hundred back to the wild, and are taking care of a hundred more.

I first met Nancy Briggs more than a decade ago when she cared for an abandoned chimpanzee in a rescue center in California. She has a heart for the great apes. As Biruté's colleague, Nancy provided help and inspiration, wrote accompanying legends, and organized photographs. She is dedicated to studying orangutan communication and saving orangutans and their habitat. As a communication scholar, Nancy has taught at California State University Long Beach for thirty years. She is a communicator par excellence. Her trips to Indonesia and Malaysia changed her life.

Karl Ammann's photographs are riveting, an amazing record of the orangutans with whom Birutée has worked so long. His many trips to Camp Leakey and surrounding areas have provided invaluable experiences for a visual journey. He partnered with expert and devoted colleagues to produce the visual excitement of *Orangutan Odyssey*.

Yet the orangutans are still the least known of the great apes to the public at large. This book provides a fascinating look at a species that deserves more attention from the world. As you read the pages that follow, I am certain you will agree that orangutans deserve all the care, support, and protection we can provide. Otherwise, the vibrant world described and preserved on these pages is in danger of vanishing forever. I know the reader will treasure this account of the orangutans as a classic.

A young chimpanzee vocalizes with Jane Goodall. (Photograph by Michael Nuegebauer.)

Chapter 1

THE MOST ENIGMATIC APE IN THE WORLD

Yet regard this wonderful monster with the human face . . . walking erect, first that young female satyr . . . hiding her face with her hands . . . weeping copiously, uttering groans, and expressing other human acts so that you would say nothing human was lacking in her but speech. The [native people] say in truth, that they can talk, but do not wish to, lest they should be compelled to labor. . . . The name they give to it is Orang outang.

Jacob Bontius (de Bond), 1658

The damp forest air is heavy with fragrance. High up in the 150-foot canopy, the crown of the giant Palaquium *tree, laced with tiny, almost invisible blossoms, embraces the sky. Seeking a glimpse of the rare and wary orangutan, I have been walking for ten days in the tropical rain forests of Tanjung Puting in central Indonesian Borneo.*

Biruté M. F. Galdikas, 1971

A female orangutan gazes into the forest where she spends most of her time. Her mother was a wild-born ex-captive. Brought to Camp Leakey, the mother mated with a wild male orangutan. Her daughter, pictured here, was born in the wild and grew up as a wild orangutan.

Borneo is part of the emerald chain of seventeen thousand islands that make up the Republic of Indonesia, strung across the equator between continental Asia and Australia. A hothouse of biodiversity, breeding countless species of wildlife, the verdant, dripping forests of Borneo and neighboring Sumatra conceal the only Asiatic great ape, the orangutan. (The gibbon resides there as well but is a lesser ape.)

Although orangutans were among the first apes known to Western science, for centuries they were the most enigmatic. Jacob Bontius, a Dutchman, made the first known reference to orangutans in the scientific literature in 1658. Compared to Borneo, Africa lies next door to Europe. When Westerners first began to explore the mountainous interior of Borneo in the late nineteenth century, the so-called Dark Continent, Africa, already had been mapped and carved into European colonies. By the mid-nineteenth century, the African apes, chimpanzees, and gorillas were familiar figures in travelogues and nature books.

Although an occasional orangutan was captured and brought to Europe, the apes were rarely observed in their natural habitat. The reason for this was simple: orangutans are almost exclusively arboreal and semisolitary. They are more difficult to find than their gregarious, ground-dwelling, and (in the case of the chimpanzee) noisy African cousins. Rarely seen or heard, orangutans evoked a body of misconceptions, and they became a mirror in which naturalists and scientists reflected the varying credos and dominant cosmologies of the age. Sometimes viewed as tool-using

buffoons for their habit of throwing twigs down at observers and for their use of objects in captivity, and at other times regarded as sage but speechless versions of ourselves, orangutans remained largely unknown not only to the outside world but also to nearby native populations.

As I walk, the raucous rasping cries of rhinoceros hornbills signal the presence of ripe fruit in the canopy. I look up to see if any other species of wildlife, whether orangutan, gibbon, or squirrel, has found the fruit advertised by their loud squawks, but the large black-and-white birds with the oversized beaks dine alone. I trudge on, my eyes scanning the canopy for any clue that might betray the presence of an orangutan.

Centuries ago, the native people of Borneo—the Dayaks, who live mainly in the interior and the Melayu, who mainly occupy the coast—recognized the orangutan's close affinity to human beings. They called the big, red apes "orang hutan," which means "people of the forest" in the Malay language, the lingua franca of the region. As Jacob Bontius reported in the opening quotation, local people in Borneo traditionally believed that orangutans could talk but chose not to because they might be enslaved and put to work.

Orangutans occasionally appear in local mythology. In a Dayak version of the "Beauty and the Beast" folktale, a male orangutan abducts a comely woman, carrying her up to his leafy nest high in the forest canopy. Although terrified at first, she finds her captor kind and gentle. After the woman has an infant, however, a child half orangutan, half human, she longs to see her parents and family. So she fashions a rope out of vines and orangutan hair she had collected from the nest, and, in the manner of Rapunzel, she flees. The orangutan pursues her. During the chase, the woman drops her infant. In his rage, the father tears the baby in half, keeping the surviving orangutan half for himself. The woman retrieves the human half and leaves the forest, returning to her village.

Thus, the Dayaks believe that orangutans and humans are so closely related that they can be sexually attracted to one another, and even procreate. Unlike Westerners, Dayaks do not draw a sharp distinction between animals and humans nor do they believe that women are the only "victims" of cross-species passion. More egalitarian than some cultures, Dayaks also tell stories of female orangutans abducting men and bearing their children. Traditionally, Dayaks believe that the connection between humans and orangutans goes back to the beginning of time. A widespread Dayak creation myth tells that God's first attempt to breathe life into humans gave birth to orangutans. Only on the second try, after God took a breather, so to speak, did humans come into being. This myth captures an essential truth and an evolutionary fact, that orangutans appeared on this planet long before modern humans did.

This Dayak myth parallels the unfolding of part of our own modern knowledge of human evolution. In 1934, a graduate student from Yale discovered fragmentary jaw remains of a hominoid (the ancestral superfamily that includes the ancestors of both humans and apes) that could be placed in the fossil record at ten to fifteen million years ago. For the next twenty to thirty years, *Ramapithecus,* as this species was called, was the oldest candidate for the first known ancestors of humans. But many paleoanthropologists were unconvinced by the fragmentary remains and questioned whether *Ramapithcecus* was even a hominid (the smaller family that includes only humans and their direct ancestors). When another paleoanthropologist, David Pilbeam, and his colleagues began excavating in the Siwalik Hills of what is now Pakistan, they hoped to verify the existence of this ancient hominid, perhaps the much sought "missing link." Instead, after examining more complete fossils, they found themselves face-to-face with what was undeniably an ancestral orangutan. The conclusion is similar to the Dayak myths. Prehumans were evolutionary latecomers, appearing only a few million years ago, after ancestral orangutans had diverged from the line leading to humans.

Ancient hominids and ancestral orangutans probably lived side by side for at least a million years. In Java, orangutanlike fossils have been found in the same layers as remains of *Homo erectus,* the hominid that preceded our own species, *Homo sapiens*. Certainly, modern orangutans and humans have interacted for thousands of years. The most compelling evidence was found in the Niah Caves of Sarawak on the northwestern coast of Borneo. Led by the late Tom Harrisson, a British scholar–adventurer who became the head of the Sarawak National Museum, archaeologists sifting through centuries of detritus on the twenty-seven-acre floor of the Great Cave, discovered a modern human skull (*Homo sapiens*) approximately forty thousand years old. With the skull were the charred bones of orangutans and other nonhuman primates. The various primate bones were exceeded in number only by the remains of wild pigs in the leftovers of these early human meals. While it is not known conclusively, the evidence suggests that the humans ate the orangutans and the pigs. Indeed, in Sarawak today, wild pigs are the game animals most favored by local Dayak hunters, and many Dayak groups still eat orangutans. As difficult as it is to accept the idea of eating creatures as close to human beings as they are, the fact is that great apes around the world have been killed for food by humans, and are still being killed today.

Two kelotocks, the wooden boats so named for the sound that their diesel engines make, arrive at the dock for Camp Leakey, where a curious orangutan sits to greet them. These boats move slowly but are the favored transportation for some two thousand people who arrive every year to see the orangutans at Camp Leakey.

Prehistoric orangutans once ranged over a much larger geographic area than they occupy today because in earlier times the climate of mainland Asia was much warmer and wetter than it is now. Orangutan homelands stretched from southern China, throughout what was once known as Indochina, into the Malay Peninsula. During the Pleistocene era, which began about three million years ago and only ended about twenty thousand years ago, the earth became colder and drier. What is usually thought of as the Ice Age was actually a long series of cold and temperate waves. With each successive ice age, much of the water in the world's oceans became locked into ice, and

Borneo's mighty rivers flow from the island's mountainous interior into the Java Sea to the south and the China Sea in the north. The major towns of Borneo are located on the coast, where the wide river mouths empty into the sea.

the sea level dropped. When this occurred in Southeast Asia land bridges emerged, connecting the islands of Borneo, Sumatra, and Java to the Asian mainland. Continental Asian wildlife migrated to the islands. Species as diverse as orangutans, tigers, tapirs, gibbons, and elephants made their appearance, only to be marooned on the islands during interglacial periods, when sea levels rose and land bridges were flooded.

Orangutans also were found on the mainland of Asia, as well as in Borneo, Sumatra, and Java. For centuries, Chinese apothecaries have sold "dragon bones" for use in traditional medicines. The dragon bones were primarily fossilized remains. Indeed, Western scientists discovered *Giantopithecus*, a large savanna ape that became extinct in China about a million years ago, not in an archaeological excavation but in local Chinese medicine shops. Ancient orangutan teeth also were found among the "dragon bones."

These ancient orangutan teeth were much larger than the equivalent of modern orangutan teeth. If tooth size is an accurate indicator of total body size, these extinct end-of-Pleistocene orangutans may have been considerably larger than modern orangutans. This created a quandary for primatologists. It seemed impossible that so large an ape could have pursued the more fully arboreal life-style of modern orangutans. According to one argument, the large, heavy ancestral orangutans must have lived in the manner of modern gorillas, traveling and foraging on the ground, with a large adult male leading and protecting the group. The gorilla analogy was popular because, like gorillas orangutans are sexually dimorphic—that is, males and females differ markedly in size and appearance. Males may weigh 300 pounds, whereas females may weigh less than 80. Mature male orangutans develop large cheekpads, a hanging throat pouch, and a beard that make them easy to distinguish from females or immature males. Typically, males and females of arboreal primates (such as gibbons) are more similar in size and appearance than are terrestrial primates. Yet orangutans are arboreal and sexually dimorphic, breaking the rules. The modern orangutan male's large size was considered by some scientists a leftover from a terrestrial life-style.

At the end of the Pleistocene era, about ten to twenty thousand years ago, orangutans became extinct on the Asian mainland. Whether they were hunted to extinction by increasingly large human populations that became armed with sophisticated pro-

Today orangutans are found only on the islands of Borneo and Sumatra.

jectile weapon systems (such as blowguns and bows and arrows), or succumbed to climatic changes is uncertain. About the same time, the orangutans also disappeared from Java and southern Sumatra, again for unknown reasons. Today, orangutans are found only on the island of Borneo and in northern Sumatra.

When I first arrived in Borneo and wandered through the great forest surrounding Camp Leakey, my first base camp on the Sekonyer River in Tanjung Puting, searching for wild orangutans, I often thought about early published images of the great apes. Like shadows or ghosts, seldom seen but often sensed, the great apes had haunted the human psyche from the dawn of Western thought. Ancient Greek mythology includes descriptions of "pygmacan races" that, like monkeys and apes, lived in the trees and were capable of making themselves invisible. The ancient Greeks had probably seen what were later called Barbary apes (not apes at all but atypical tail-less monkeys) which roamed the Mediterranean shores of North Africa, as well as the monkeys kept by their Egyptian contemporaries, who revered the male hamadryas baboon, with his majestic silvery, lionine mane. Egyptians believed that the male hamadryas was an incarnation of Thoth, the god of scribes and scholars and the inventor of science and writing, who stood behind the king of gods in divine assemblies, inspiring wisdom.

The first record of the actual observation of great apes occurred relatively early. In 470 B.C. a group of colonists sailed down the coast of West Africa, where they encountered hairy, stone-throwing creatures they called gorillai (probably a local name). What they were remains unknown, but in any case two thousand years would pass before Europeans began to learn more about what we now know as gorillas.

During Roman times, Pliny the Elder described all manner of exotic, half-human creatures, including one race that hopped on one foot like an invention of Dr. Seuss. Pliny's fantasia included the "satyrs" of India, who had the legs of a goat and the upper body of a monkey or human, and creatures with dog-shaped heads and furry clothing, that lived in the mountains. The "satyrs" might have been based on the lanky hanuman langurs, believed by Hindus to be manifestations of Hanuman the Monkey God. The different species of macaques that reside in India may have inspired tales of the dog-faced mountain creatures.

In the fifteenth and sixteenth centuries, the world was opened up to exploration by Portuguese navigators in small wooden sailing ships able to sail into the wind, a new technique that for the first time allowed European explorers and adventurers to return home relatively quickly after their voyages of discovery. Previously, they could return home in square-riggers that sailed "before" the wind, but the journeys were very long, and the routes were circuitous. Suddenly, Europe was flooded not only with the gold and silver of conquered New World empires, trade goods from India and China, and gold, ivory, and (later) slaves from the west coast of Africa but also with plant and animal specimens from the newly discovered tropical world. The recent invention of the printing press helped to spread explorers' tales of exotic places flora and fauna. Among the specimens of the fauna brought to Europe for the first time, alive or dead, were great apes.

In 1641 a Dutch anatomist, Nicolaes Tulp (coincidentally the subject of one of Rembrandt's greatest paintings), provided the first detailed description of an ape. The animal had been given to the Prince of Orange, ruler of the Netherlands. Tulp classified the ape *Satyrus indicus*, harking back to Pliny's "satyr," but noted that the animal was called "orang-outing" by natives in the ape's land of origin. In retrospect, Tulp's description suggests that the ape was not an orangutan but more likely a bonobo (or pygmy chimpanzee).

Half a century later, Edward Tyson dissected the first ape to reach England alive, an infant who, unfortunately, died soon after arrival. Tyson also called this ape "orange-outing," even though the infant was almost certainly a chimpanzee. For more than a century, variations of the Malay name for orangutans were applied to all apes, African and Asian. Why this was so remains a puzzle.

Curiosity and wonder about apes also were fueled by travelers' accounts and stories collected from sailors, soldiers, and sea captains. A mix of direct observations, descriptions, and myths learned from local peoples, and just plain imagination, such accounts in book form were among the best-sellers of their day. In 1607, Andrew Battell, an Englishman who had been imprisoned by the Portuguese in West Africa, returned to England. His memoir, published in 1625, included what was probably the earliest description of gorillas and chimpanzees as "two kinds of Monsters, which are common and very dangerous." Shakespeare's Caliban, a hybrid of human and beast, may have been inspired by Battell's monsters.

The first Englishman who visited orangutans in their native islands was Daniel Beeckman, a ship's captain. In his 1714 book, *A Voyage to and from the Island of Borneo,* he described "oran-ootans" as having "larger arms than men [and] tolerable good Faces." He continued, "nimble footed and mighty strong; they throw great stones . . . [and] sticks at those persons that offend them." For the time it was a surprisingly accurate and objective picture. Unfortunately, Beeckman's pet orangutan infant died after only seven months in captivity and so was never seen by other Europeans.

In the eighteenth and early nineteenth centuries, Westerners struggled to sort out the real from the imaginary, sometimes unsuccessfully, in their classifications of the living world. At this point, the specimens available for the study were chimpanzees from West Africa and orangutans from the Dutch East Indies. Europeans had heard of gorillas from the local people but had not seen or captured one.

The great French naturalist Comte Georges-Louis de Buffon concluded, on the basis of virtually no evidence, that there was only one species of ape, which varied in size. He called the small ones (including Tulp's carefully described bonobo and Tyson's chimpanzee) "Jockos" and the large ones (orangutans and the legendary gorillas) "Pongos." Others classified apes by color. In this view, there were two types of orangutans, a black one from Africa (chimpanzees and bonobos) and a red one from Asia. The Dutch anatomist Pieter Camper, the first scientist to conduct an intensive study of ape variability, concluded that orangutans and chimpanzees were different species, providing orangutans with their own identity.

Even so, orangutans confused scientists. The Dayak and Melayu people of Borneo believed that adult male orangutans, with their enormous size, bulging throat pouches, and face-expanding cheekpads, and the smaller female and subadult orangutans, were different types. Likewise, Western naturalists assumed that several species of orangutans existed.

The publication of Charles Darwin's book *On the Origin of Species by Means of Natural Selection* in 1859 electrified the Western world. The book contained only one sentence suggesting that the theory of evolution might shed light on human origins, but that hint (elaborated on in *The Descent of Man* in 1871) was enough. Although Darwin did not in his first book suggest any such thing, the idea that human beings were descended from apelike ancestors rekindled fascination with the apes and a market for specimens. Alfred Russell Wallace, the codiscoverer with Darwin of the principle of natural selection, had shot dozens of orangutans while collecting flora and fauna in the Far East. Equally industrious, if not more so, was William Hornaday, who sold natural history specimens to schools, museums, and other scientific institutions. On one day alone he shot seven orangutans, including one that he had found innocently peering down from a nest.

Ironically, and perhaps paradoxically, Darwin's perception that apes are our closest living relatives led not to a family embrace but rather to their further slaughter and vilification. The great eighteenth-century Swedish botanist, Carolus Linnaeus, had grouped living creatures according to their physical similarities, and he postulated a hierarchy of beings from the simplest to the most complex, culminating with humans. (His taxonomy begins with the largest grouping and continues to the smallest—Kingdom, Phylum, Class, Order, Family, Genus, Species.) He put monkeys, apes, and humans in the same order, Primates (or first). Like most of his contemporaries, Linnaeus believed that this Great Chain of Being was the result of divine creation and was, therefore, perfect and immutable. What else could explain the orderly diversity of life on earth? But with the Darwinian revolution, humankind's inexorably superior position in the universe was no longer assured by the act of special creation through which an omnipotent deity had birthed the human species. A new means had to be found to draw a line between ourselves and the beasts. Related to us by blood (as it were) and by descent, apes were the threat closest to home. To demonstrate human preeminence, most philosophers, priests, scientists, and artists were forced to repudiate and vilify their nearest of kin.

From the nineteenth century onward, the image of apes in Western culture became increasingly savage and brutal. Edgar Allan Poe used this new monster ape image in "The Murders in the Rue Morgue" (1847), the first modern detective story. He described the villain who brutally killed two women as a creature "of an agility astounding, a strength superhuman, a ferocity brutal, a butchery without motive, a grotesquerie in horror absolutely alien from humanity." In Poe's story, the identity of the brute (a serial killer in today's terminology) is revealed when the finger marks on the throat of one of the victims perfectly matches the measurements of an orangutan's hand.

This monster image reappeared in the tales of the American explorer and journalist Paul Du Chaillu, who visited Africa in the 1850s. Du Chaillu thrilled readers with his accounts of charging adult male gorillas, which he slaughtered by the score.

In the early morning mist, the author and Pak Bohap walk with the orangutan Siswi along the ironwood bridge under the landmark lookout tower of Camp Leakey. The two-hundred-yard-long ironwood bridge over the tropical peat swamp forest connects Camp Leakey to the Sekonyer Kanan River, a tributary of the Sekonyer River. The Sekonyer is a classic black-water river, so called because the dark river reflects the surroundings and the sky, preventing the human eye from penetrating its depths.

An orangutan infant, leisurely hanging from a low tree over the Sekonyer River, chews on a fruit. This infant, about three years old, will stay close to his mother and will cling to her when she moves from tree to tree in the canopy. Immature orangutans may remain with their mothers until they are nine or ten years of age.

A traditional belief of the indigenous Dayak and Melayu people of Borneo was that adult male orangutans are actually ghosts. Here, a mature old male, with distinctive cheekpads, lies down for a rest in the protection of the Tanjung Puting National Park. The survival of his species is threatened by the diminishment of the Borneo and Sumatra tropical rain forests due to fire, clear-cutting for lumber, and the establishment of palm oil plantations.

Gorillas, he wrote, were "some hellish dream-creature—a being of that hideous order, half-man, half-beast," confirming a nightmare version of the ape as a fiend straight from the depths of Hades, a degraded form of life, a creature whose failure to attain human status was a sign of failure in the evolutionary scheme. Another variation of this brutal ape image appeared in 1912 with the publication of Edgar Rice Burroughs's *Tarzan of the Apes*:

> The ape was a great bull weighing probably three hundred pounds. His nasty, close-set eyes gleamed hatred from beneath his shaggy brows, while his great canine fangs were bared in a horrible snarl as he paused a moment before his prey. . . . With a wild scream, he was upon her, tearing a great piece from her side with his mighty teeth, and striking her viciously upon her head and shoulders with a broken tree limb until her skull was crushed to a jelly. . . . Standing erect he threw his head far back and looking fully into the eye of the rising moon he beat upon his breast with his great hairy paws and emitted his fearful roaring shriek.

Great apes got a brief respite from vilification in the mythmaking capital of the modern world, Hollywood, when a series of Tarzan films showed the hero becoming pals with a benign juvenile chimpanzee. But the monster-ape reappeared in the enormously influential film *King Kong* (1933), which drew bigger audiences than any other film produced that year and became a classic. Most closely resembling a gorilla, the ape in the film nevertheless caused orangutans and chimpanzees to be associated with this terrifying new vision of apes. Even today, Indonesians and Malaysians sometimes refer to orangutans as King Kong, demonstrating the power that Hollywood exerts in the shaping of perceptions of the natural world. In reality, gorillas are usually peaceful vegetarians, and the adult male silverbacks who terrorized movie audiences are exceptionally tender fathers.

A new perception of the personality and character of the great apes only occurred near the middle of the twentieth century. The end of World War II brought unprecedented prosperity to North America, and with Europe still recovering from the ravages of war and many African and Asian nations busy tossing off the fetters of colonialism, the United States and Canada entered an era of calm, domesticity, easy employment for most people, and a car in every garage. By the late 1950s and early 1960s a new generation had declared its opposition to war and its desire to return to nature and simplicity. "Flower power" was a catchphrase, and "make love, not war" became a rallying cry of a generation opposed to the aggressive materialism of society and the conventionality of the parental generation.

The ape to benefit most from the preoccupations of the new generation was the chimpanzee, which is, ironically, the male ape most prone to intercommunity aggression, infanticide, and predation. Nevertheless, Jane Goodall's and the British primatologist Vernon Reynolds's first reports of gentle, peaceable (albeit noisy) chimpanzees living in "open societies" coincided and supported the perceptions of the age. The idyllic "flower child" image of chimpanzees in Eden mirrored the times. This image was enhanced by Jane Goodall's pioneering discovery that chimpanzees both used and made tools. The new generation cheered the news that chimps were, after all, "just like us," or how we wanted to be, once again in Eden.

Gorillas, the greatest of the great apes, with adult males in the wild weighing more than four hundred pounds, also benefited from the attitudes of the new generation. Gorillas, it was learned, rarely attack humans except in self-defense. It was further reassuring to discover that the huge silverbacks will die to protect their families. These gentle giants became much more appreciated as a result of the work of Dian Fossey, whose portraits of individual gorillas, etched in sympathy and loving detail, began appearing in *National Geographic* magazine, which also had published Jane Goodall's work for a very large popular audience.

By this time, the dramatically announced discoveries of prehuman species (*Zinjanthropus, Homo habilis*) by Louis and Mary Leakey in East Africa had cemented the new generation's visceral acceptance of human evolution as a process from which modern humans had eventually emerged from apelike ancestors. To learn more about these early prehuman ancestors, attention centered on the chimpanzee, humankind's closest living relative in the animal kingdom, sharing almost 99 percent of its genetic material with humans. The chimpanzee became a model for the creature, known only from incomplete fossil skeletons, from which humans descended. Louis Leakey encouraged Jane Goodall to study chimpanzees at Gombe Stream Reserve (now a national park) as a way of extrapolating from chimpanzee behavior the life-style of the ancient hominids whose remains he and Mary Leakey had found in the dry dust of Olduvai Gorge. Jane Goodall's *National Geographic* magazine articles and books, combined with Hugo van Lawick's films, brought chimpanzees to the forefront of American consciousness. The lush settings of Gombe, low hills with waterfalls in their bosoms, nestling against the sparkling waters of an azure African lake, encouraged expectations that this was, indeed, Eden and that the chimpanzees were almost ancestors. Such well-observed behavior as chimpanzee aggression, predation, and infanticide passed over the heads of the general public.

Orangutans now also benefited from this new benign image of the great apes. But an aura of mystery remained securely attached to these apes because most researchers flooded to Africa to study primates. (Gombe became a bustling international research center.) In the early 1960s, however, some progress was made. Barbara Harrisson, the wife of the then head of the Sarawak Museum, Tom Harrisson, published her charming book *Orangutan.* Barbara Harrisson had rescued and raised orphaned infant orangutans, whom she described in detail, and some of whom entered the first orangutan rehabilitation program for return to the wild. But her fleeting encounters with wild orangutans only added to the mystery. Even the extraordinarily skilled field biologist George Schaller failed to add much to our knowledge. After spending a year with mountain gorillas in Africa, Schaller observed wild orangutans in Sarawak on the northwest coast of Borneo. Despite considerable effort, his report primarily enlightened people on the difficulty of studying orangutans in the wild, and published accounts by several other scientists had the same effect.

Unyuk, an adult female orangutan at Camp Leakey, walks upright, with her infant clinging to her back like a papoose. Mother and child hold pieces of fruit. Female orangutans share food with their young, which is how the young learn about the four hundred different food types that wild orangutans consume in the forest. Unyuk's bipedal walking on the ground is unique, although an upright stance often occurs among wild orangutans who stand in the canopy holding onto adjacent branches.

Wild orangutans can be observed but it is infinitely more difficult to photograph them. The mechanical attributes of the camera and the light-grabbing sensitivity of the film do not approach the effectiveness of the human eye. Orangutans spend most of their time high up in the tree canopy, moving frequently during the day, and light conditions for photography up there are next to impossible. Even if one could stake out a photographic blind, with lights to aid the poor visibility, the chances of a wild orangutan coming in range are remote. Most of the photographs in this book, therefore, are of the more accessible wild-born, free-ranging, ex-captive orangutans. Still, the behavior of the animals is virtually identical. The scenarios that the reader will see played out in these photographs among wild-born ex-captive orangutan mothers who had been released earlier as juveniles or adolescents at Camp Leakey are replicated by wild orangutan mothers and infants.

During the 1960s, as Americans went to the moon, the orangutan resolutely remained unknown. The behavior and life cycle of orangutans in captivity were reasonably understood by then, but what was known about wild orangutans barely filled a page. These were the facts in hand: the animals were arboreal, with curiously expressive faces; males produced very loud "burps," perhaps as a warning call; and orangutans were highly endangered and likely to disappear. Estimates of their numbers hovered at fewer than five thousand.

When I entered the field in 1971, encouraged by Louis Leakey like Jane Goodall and Dian Fossey before me, three Western scientists had just successfully completed some two years of studies of orangutans in North Borneo (Sabah) and East Borneo. Their studies were unavailable to me, however, not yet having been published. When I arrived in Kalimantan Tengah (Central Indonesian Borneo), the mystery surrounding orangutans was so great that it was not known whether they were social or solitary; whether they were herbivorous or frugivorous; and whether they were totally arboreal or occasionally came down to the ground. Some scientists assumed that orangutans must be ginger-haired Asian chimpanzees; others saw them as something utterly alien in their reclusiveness and solitary behavior. Orangutans were still the great unknown. When I first went to Borneo, I had little idea of what I would actually find.

I trudge on. After walking almost eleven hours today without encountering even one wild orangutan, I am exhausted, sweat dripping from every pore, my tired feet flaming with soreness. Suddenly in the distance I hear branches cracking. The sound is repeated. I race to the spot. There in the canopy, at least one hundred feet above the forest floor, an adult female orangutan with an infant tucked under her arm is leisurely constructing a nest. It is still an hour before dark.

I watch as the female exhibits engineering skills denied me. After bending branches to form a sturdy platform, she lines the nest with sturdy leafy branches to form a springy bushy orangutan version of a mattress. She lies down on her back, her infant on her stomach, one foot hooked around a branch outside the nest. Contented with her day's foraging on flowers and fruit, she turns in her nest. Contented now finally with contacting a wild orangutan, I turn and leave. Tomorrow I will return at dawn to begin my study.

OPPOSITE: ***Two juvenile orangutans play-fight in the lower tree canopy. Acrobats of the rain forest, orangutans move with ease through the trees. Anatomically, with their short trunks, long arms, and hooklike hands with long fingers, orangutans are suited for hand-over-hand movement, called brachiating. In reality, they rarely move that way, preferring to clamber and climb cautiously through the canopy using all four limbs.***

The position of the orangutan's thumb on the hand is a structural adaptation for the hooklike grip necessary for moving hand over hand in the trees. When humans brachiate, their thumbs get in the way. Below, photographer Karl Ammann holds hands with an orangutan in a mutual gesture of friendship and respect. Because of their relatively close evolutionary relationship, humans and orangutans readily understand each other's gestures.

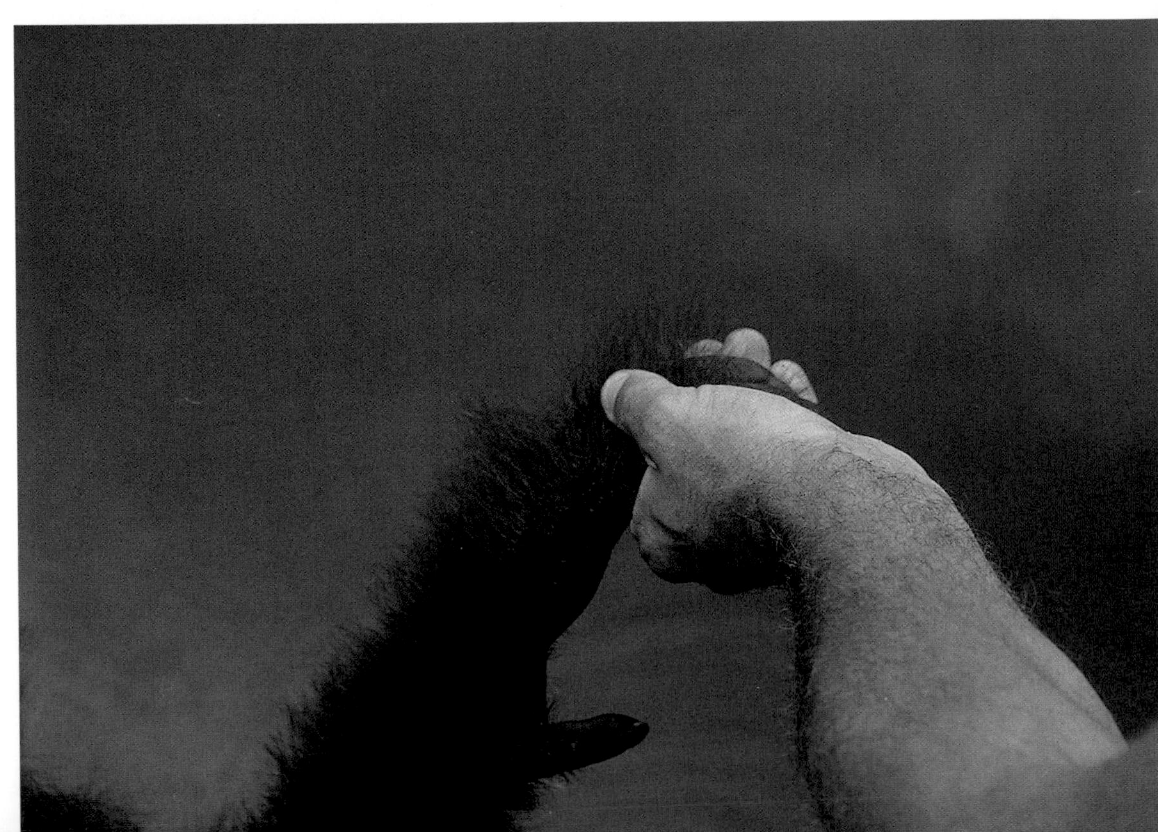

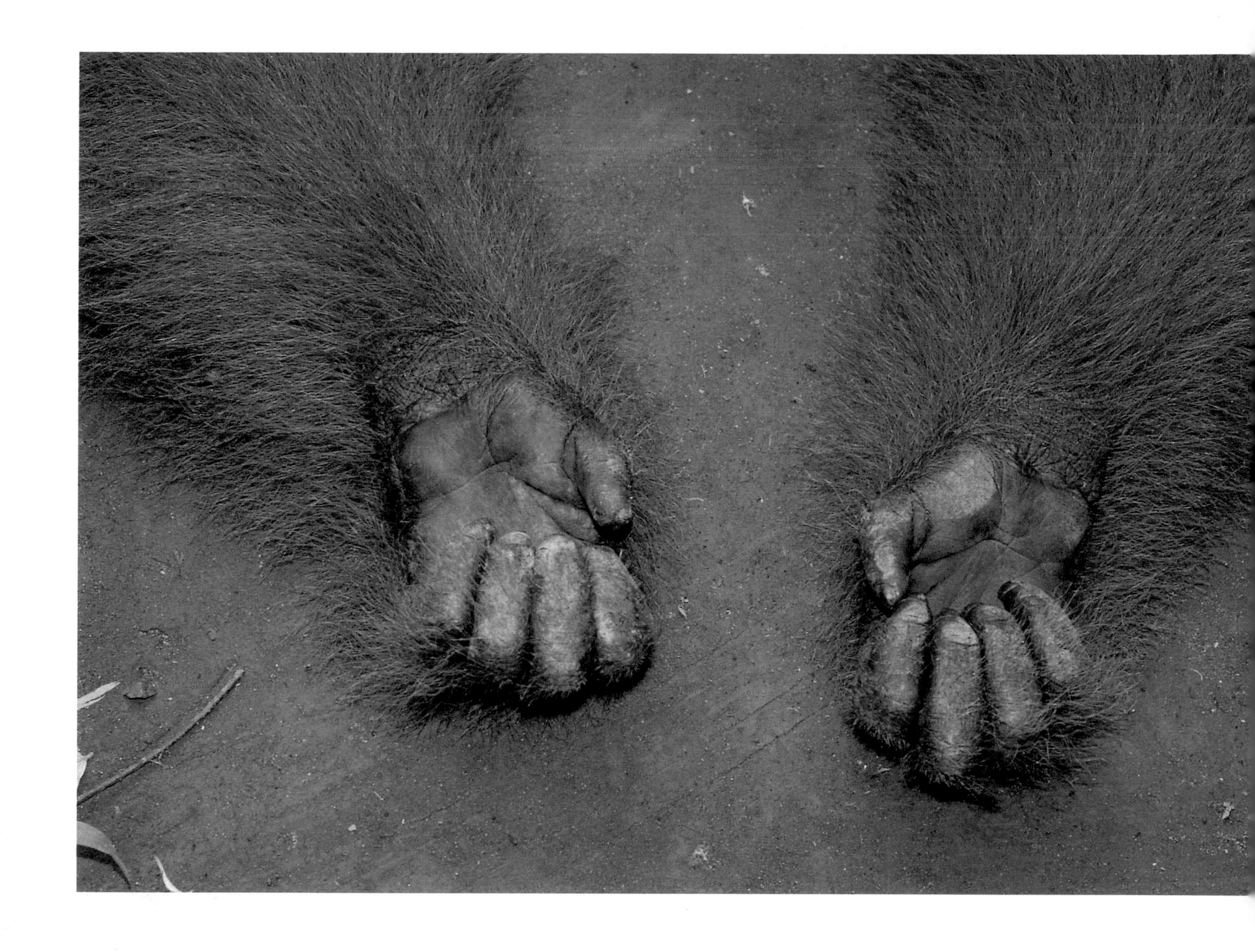

With arms outstretched above his head, an orangutan relaxes on the ground as though he were sunbathing. The notable calluses on his hands give evidence of the extensive use he makes of them in walking on the ground and in reaching, grabbing, picking, scraping, prying, and opening fruit and other items from rough branches.

OPPOSITE: *A playful juvenile demonstrates the extraordinary strength of the orangutan grip as he happily swings from a vine on one arm.*

All four extremities are useful from time to time. Orangutan feet are used like hands.

Huddling close together for comfort in a canoe, two wild-born, ex-captive orangutans demonstrate their need for a mother's love. Orangutan mothers in the wild suckle their juveniles until they are at least five or six years old, and nursing can go on for seven or eight years.

The author walks through a grassy field near Camp Leakey with some of the wild-born, ex-captive orangutans that she was beginning then, in 1975, to rehabilitate to life in the wild. (Photographs at left and above by Rod Brindamour.)

A 1980 cover of National Geographic *magazine featured the author's first child, one-year-old Binti Brindamour, and a wild-born, ex-captive infant female, Princess. The young female was being bathed when Binti, fully clothed, jumped into the tub. Princess at once playfully climbed onto Binti's back. (Photograph by Rod Brindamour.)*

From 1978 to 1980, Gary Shapiro, then a graduate student from the University of Oklahoma, taught American Sign Language gestures to Rinnie, a free-ranging wild-born, ex-captive adult female orangutan. LEFT: *At one session, the author helped record the process, though somewhat impeded by other friendly orangutans. Still devoted to the study and preservation of orangutans, Gary Shapiro is now Vice President of Orangutan Foundation International. (Photograph by Rod Brindamour.)*

A distant view of an orangutan nest at the top of the tree canopy shows the difficulty of observing wild orangutans. The difficulty is, of course, compounded many times over when the observer is inside the tropical rain forest, not on the edge of a clearing, as seen here.

Chapter 2

THE BEST MOTHERS IN THE WORLD

No one like one's mother ever lived.
Robert Lowell

The infant orangutan's buttery chocolate eyes peer out from under his mother's armpit. He is nestled so comfortably on her body that he seems joined to her at the hip. Mother and son are one. Just as dawn is breaking, the wild mother and infant leave their night nest. Down below it is still dark, but up in the canopy, glimmers of light are visible. The orangutan is feeding in a tropical oak tree (Lithocarpus). *High up at the top of the canopy she reaches out for large flat white acorns, which she gathers in the one hand. After putting them in her mouth, she noisily crunches on the acorns, spitting out bits and pieces. The infant is as placid as an infant can be. He is approximately one year old. He stares off into space in that peculiar oblivious and unself-conscious way that is typical of wild orangutans of all ages. If the word contentment were a picture, he would appear in the dictionary representing the concept.*

As I learned over the years, many adult females stay in the same general area of forest, in a range of about ten square miles, whereas most males come and go. Thus, because of the way orangutans live, one focus of my study became adult females. I observed wild adult females moving rapidly through the canopy on "trails" of vines, branches, and flexible pole trees, whereas in the swampy terrain below, the human observer struggles to move one foot at a time through the sucking mud and long slimy roots. Semisolitary and mostly silent, the female does not announce the discovery of a fruiting tree, as the more gregarious chimpanzees do. When she stops to eat or rest, she disappears into the foliage. Only the sound of snapping twigs and champing betrays her presence. Now and then a shaft of light sets her red-orange coat ablaze, but out in the sun her dark skin hides her so well that she is virtually invisible. It is not surprising that George Schaller and other investigators had such difficulty observing wild orangutans, male or female.

I found that I could maintain contact with female orangutans only by waiting below as one painstakingly built a nest at dusk and then returning to the same place before dawn broke and the female began moving again. Through many thousands of hours of observation spanning almost three decades, I gradually obtained a detailed picture of wild orangutan mothers and their offspring.

Multiple births are common among many animals. Interestingly, multiple births occur at about the same low rate for captive orangutans as for humans. While no wild orangutan twins have been observed, about one out of ninety natural pregnancies

The face of a juvenile orangutan.

produces twins among captive orangutans, the same percentage as with humans. As for frequency of birth, among the mammals, many of larger size, where singletons are the norm, the mother bears offspring as often as once every one or two years. The orangutan is different. As I eventually discovered, no other wild mammal gives birth so rarely. A wild orangutan female at Tanjung Puting will, on average, give birth only once every eight years.

Orangutan pregnancy lasts about eight months, approaching the gestation time of humans. Compared to other animals, orangutans grow very slowly. Indeed, compared to other nonhuman primates, all the great apes are developmental laggards in this regard. Perhaps a long period of growth is required because the great apes have much to learn and need a long dependency (or childhood) in which to learn it.

Most nonhuman primate mothers carry their offspring on their bodies after they are born. Nowhere is this attachment so intense and long-lived as among wild orangutans. For the first six months of life, the infant is literally glued to the mother. While the youngster may crawl over and around the mother's body, he or she never leaves it except perhaps when the mother lies down in her night nest. The infant may lie down touching her side. Then during the first midyear of life, the little one takes the first tentative steps along a branch.

By the end of the first year, the infant seems comfortable in the trees a few feet away from the mother, but still seems weak-kneed and unbalanced, appearing to be most at ease hanging suspended from a twig. By the end of the second year, an infant's motor skills have markedly improved, although its weight and height have increased relatively little. The infant becomes quite agile in the trees, able to brachiate and dangle by one arm. It begins to show interest in manipulating vegetation. By the third and fourth years, the infant has entered the equivalent of the human toddler's "terrible twos." Although it still clings to its mother when she moves, the moment she stops the infant is off. The youngster will rarely leave the crown of the tree where the mother is sitting or feeding but will explore or play in her vicinity.

The adult female orangutan continues to carry her offspring for the first four years of life as she moves from tree to tree. Not until the offspring's fifth year will he or she begin to travel behind the mother. Since the crown of the rain forest emergent can stretch up to eighty feet wide, I was sometimes alarmed when I looked up and easily found a mother but could not spot her infant anywhere. Something, I thought, must have happened to the infant, something tragic for one so young not to be in the same tree with the mother. After peering into the treetops anxiously and circling around and around the tree repeatedly to get a better view, I would eventually see the tiny infant on a twig in the uppermost reaches of the tree or industriously exploring the complexities of a particular branch, almost hidden by the foliage.

Wild orangutan infancy is probably the least stressful and most idyllic of any childhood in the tropical rain forest. First and foremost, an orangutan infant does not have to share a mother's attention with any other creature. Although an older offspring may be traveling with mother and baby, the juvenile or young adolescent is usually some distance behind, does not nurse, and does not share the night nest with the rest of the family. As far as the infant is concerned, the older sibling is merely an occasional diversion and perhaps also a welcome playmate. Nor does the infant share a mother's attention with her mate, because the adult male orangutans do not travel with females or the young.

Among orangutans, males and females lead separate lives, and single mother hood is the norm. Around the time of conception, the mature male who copulated with the female may spend several months, off and on, traveling with the mother-to-be (and chasing off other males). After this brief honeymoon, the male moves on in search of other females. In several cases I suspected that the same male returned years later when a female whom he had made pregnant was ready to mate again, but except for these instances I have observed no lasting parental or other bond between orangutan females and males.

On occasion, a subadult male may attempt copulation with a mothering female, but the female is usually able to evade him. She also avoids much contact with adult males. Occasionally, an adult male will approach a mother and infant, but this interlude is brief. Sometimes while following a mother and infant, I glimpsed an older male, a cheekpadder, in the distance. Most often, mature males make their presence known with their majestic long calls but are rarely seen.

Usually, an infant is alone with its mother, cradled as if in a cocoon by her large powerful body. The infant could not be more snug in a kangaroo mother's pouch. The young one suckles on demand. High up in the tropical rain forest canopy, the large adult female orangutan fears no natural predator. The infant, protected by mother's invariable presence, appears to have the most secure place in the universe. Occasionally, while an infant is playing on a twig or climbing among branches as a human child would on a jungle gym, the mother begins to leave. To an infant orangutan, mother is the center of the world. The high-pitched squealing and grunting that the mother's unanticipated departure provokes are among the most heartrending cries imaginable. A human baby's wail simply does not compare. The almost total silence of orangutans in the forest is such that the infant's squealing seems as jarring as a fire siren.

The infant not only suckles on demand but also takes food from the mother's mouth, hands, and, sometimes, even from her prehensile feet. Many foods eaten by orangutans are difficult to process. Fruit contents are sometimes protected by thick horny skins, spines, burrs, and even burn-causing or sticky organic substances; barks are hard to peel. While a mother is working to prepare food for eating, the infant frequently wanders off in the tree, plays, forages a bit, and then comes back to the adult. The youngster sticks its face next to the mother's mouth, watching her intently in the classic begging gesture typical of great apes. (I have seen exactly the same intent expression on a chimpanzee's face in a film by Hugo van Lawick. The chimpanzee female waited for an adult male to spit out a piece of gristle from his mouth. The male had just killed a red colobus monkey and was surrounded by an admiring throng of fellow chimpanzees who hoped he would share his kill.) When an infant orangutan approaches a mother in this way, the adult usually (but not always) allows the young one to take whatever he or she wants. Sometimes the female spits the chewed remains of the processed food into her infant's hand or allows the little one a taste of the mush in her mouth. An instance comes to mind: during my first multiple-day follow of an

adult female that I called Beth and her son, Bert, I noticed that she was feeding on some bark, which she was holding in her hands and feet. Bert was still probably only two years old. Bert disappeared for a few seconds from my view. When he reemerged, I noticed that he, too, was holding a piece of bark. For a minute or two, I was befuddled. Beth had not moved a muscle and was still holding bark in her hands and feet. Yet there was no way that the infant could have scraped a piece of bark from the tree in which Beth was feeding during the few seconds that he was out of view. He must have gotten the bark from Beth, perhaps detaching a piece from the material in her hand, or taking it directly from her mouth. It didn't matter how he got it, the point was that he got it.

Modern primatologists have always emphasized how dependent nonhuman primates, especially monkeys and apes, are on learning. Infant primates must learn not only what to eat but also how to obtain and process their food. Orangutan infants learn from their mothers. At about six months of age, very much like human babies, infant orangutans begin to sample the solid foods consumed by their mothers by taking food from their mouths. The process is gradual and takes place in fits and starts. When it begins, the mother does not invariably share her food. Sometimes she turns her head away just as the infant reaches for her mouth. At other times she may move her hand just as the infant grasps for the food tidbit she is holding. Denial of the food frequently results in violent temper tantrums, during which the infant is momentarily transformed into a screaming spoiled brat, instantly recognizable to human parents who refuse to buy candy or chewing gum for their offspring at a grocery checkout counter. After a display that the parent of any any human toddler would recognize, the infant orangutan returns to its mother, who now relents and calmly allows her infant a bit of food as though absolutely nothing had happened. By the time an infant is three or four years of age, the squealing that occurs when a mother orangutan does not wish to share her fruit reaches a peak of intensity. Most of the time, however, mothers do share food with their offspring, but in such a seemingly careless and oblivious manner that it seems almost as if she were unaware of what she is doing.

A mature wild orangutan mother stares impassively out of her perch in a tree with her infant at her shoulder. Wild orangutans like this are difficult to study, but in addition to the considerable work with wild-born, ex-captive orangutans, the multigenerational study of wild orangutans is now approaching three decades of work based at Camp Leakey.

Infant orangutans learn about nutrition from examining the spit out husks or skins and inedible seeds of the fruit that adults consume, much as I learned about the orangutan diet by collecting the litter that rained down on me from the tree canopy. How better for a youngster to find out what part of the fruit, leaf, or bark is edible than to study the leavings and take what is edible directly, as it were, from the horse's mouth, in this case, mother orangutan.

Some anthropologists have suggested that food sharing between a mother and her offspring buffers the weaning process, allowing the dependent youngster access to a new source of solid food while the supply of milk is diminishing. The intense sharing that I have seen taking place between mother orangutans and their offspring may buffer the juvenile during weaning, but the function of the sharing is far deeper. Infant orangutans frequently play with the food. Also, the pieces of the food they get from the mothers are not always of the highest nutritional quality. So it is almost as if the mother orangutans know that the infants are, in effect, practicing foraging: plucking, peeling, sniffing, and tasting the fruit.

A one-year-old infant puts his mouth to his mother's lips as if to beg for food and in playful affection.

OPPOSITE: *An infant clings anxiously to his mother.*

Sharing food, a mother dribbles morsels out into a youngster's hand.

The color of hair or skin of many nonhuman infant primates changes as they grow older. Baboon infants are born with pink faces and ears and black coats of hair that change to an adult coloration before the infant baboons are a year old. Chimpanzee youngsters display a tuft of white hair on their behinds that eventually disappears at about the age of one. Among the growth changes that occur in orangutans is the color of their hair. Humans are no exception. My oldest son was platinum blond as a toddler. Now, as a young adult, his hair is dark blond. An orangutan infant's hair may be of lighter hue than its mother's hair. The greatest difference, however, between orangutan adults and infants is the prominent white patches around an infant's eye and mouth as well as scattered over its body. With some infants the effect is very subtle. With others the effect almost matches that of piebald ponies. As the infant ages, the patches become darker, and eventually, at adulthood, the orangutan's face and skin are almost black beneath the sometimes sparse orange hair. The darkness of the facial patches is a rough but powerful gauge of an orangutan's age.

Another mother gingerly grooms her child with her fingers. Although the orangutan hand is considerably different from the human hand, the thumb is opposable, making delicate gestures possible.

Some of the ex-captive mothers that I have worked with have occasionally returned to camp, frequently for supplemental feedings but also for an interlude of human companionship and activity. These orangutans were fully rehabilitated, and were not in any way to be treated or thought of as pets. The orangutans invariably returned to the forest, but while they were at or near Camp Leakey I was able to watch and even interact with them remarkably closeup. The wild-born, ex-captive mothers enhanced our understanding of orangutan parenthood by enabling us to see the details better. But it was the wild orangutans who provided the bedrock of our understanding.

Among the orangutan mothers we got to know especially well at Camp Leakey were Unyuk, Princess, and Kuspati. As youngsters they had been released at Camp Leakey at different times. Princess had a distinctive early experience. After her arrival at Camp Leakey as a small juvenile she was taught a sign language similar to that used among deaf humans by Gary Shapiro, then a graduate student and currently vice president of the Orangutan Foundation International. In the nearly two years that Shapiro was a surrogate parent to Princess, he taught her more than thirty signs, which she used individually or combined to create short phrases or sentences. Like a good orangutan parent, Shapiro carried Princess everywhere. In the evening he would lay her down to sleep in her own room in the simple wooden building that was his home at Camp Leakey.

The three orangutans differ significantly in personality. Princess is shrewd, gentle, and seems to enjoy the company of humans. Unyuk, having arrived at Camp Leakey as a juvenile, is quick, outgoing, and assertive. As the eldest, she is the highest ranking of the three. Kuspati was reared in the forest, as opposed to being hand reared (like Princess), and is shy and more wary of people. Once past infancy, Kuspati spent much of her time in the forest with other juvenile orangutan peers.

Unyuk, Princess, and Kuspati are ex-captive orangutans who became mothers after their release at Camp Leakey. All three were extremely tender with their infants. When the infants started to move away, the mothers did not impetuously grab them but gently reached out to bring them back onto their bodies. Although relatively comfortable around humans and with each other, these mothers did not allow their older offspring or another orangutan mother, much less a human, to handle their infants.

Orangutans rarely groom one another. Mothers are an exception, although grooming bouts with infants usually last only a few minutes. Yet at Camp Leakey, the sweet care with which these mothers picked through the hair of their infants was heartening. Anyone who ever watched Unyuk, Princess, or other adult females came away with admiration for their devotion and skill as mothers.

Wild-born, ex-captive females spend more time on the ground than do wild orangutans. They also have access to supplemental feedings. As a result, their offspring seem more independent than juveniles and infants in the wild. Probably because of the feedings, infants of ex-captive mothers were usually larger than wild infants of the same age (although smaller than captive infants in North American zoos). In addition, the infants and juveniles who visited Camp Leakey with their mothers shared more social interactions and play sessions with other young orangutans than wild orangutans do. The offspring of ex-captives seem more socialized, at least during their youth. This may be an advantage for their survival in the wild.

Because they lead semisolitary lives, wild orangutans do not need, or do not display, the social skills evident in developing chimpanzees and gorillas, who live in groups. Observations at Camp Leakey showed that when orangutans encounter other individuals, these social skills emerge quickly and relatively effortlessly. When brought together by food, female orangutans demonstrate a respect for rank that minimizes conflict, and infant orangutans display a delight in social play equal to that of the other great apes. Wild orangutans do not usually make tools, as chimpanzees do, but when exposed to humans and their possessions, orangutans quickly learn not only to attend to human activities and artifacts but also to imitate humans and to use objects in novel and creative ways for their own purposes. Orangutans are quick students: at Camp Leakey they have been observed taking dugout canoes to get to the other side of the river and sometimes seemingly for joyrides. One subadult male named Gundul even used a dugout canoe for a portable river-edge nest that he would drag behind him with a rope.

Unyuk, Princess, and Kuspati were all excellent mothers, but their parenting styles, like their personalities, differed. (I also observed this variability among wild orangutans.) One of the main differences I noticed was how much they shared with their offspring. Unyuk, though tough and boisterous, shared food generously with her oldest offspring, Uranus, even after his younger brother, Udik, was born. (Unyuk's open spirit of generosity did not extend to people—she always expected people to share their food with her.) Princess was less generous with Prince, her first offspring. Once Peta, her next infant, began to taste solid food, around the age of one, Princess virtually never shared with Prince. The young male begged, screamed, and threw temper tantrums. Princess sometimes responded by snapping, after which the squealing Prince would move away.

Unyuk also engaged in a great deal of sex play with her son Uranus, unlike Princess and Kuspati, who remained more aloof. It was just an expression of their different personalities. If their infant sons masturbated on their mothers' toes, fingers, or various orifices (sometimes almost hilariously inappropriately), Princess or Kuspati gently swatted them. Unyuk, in contrast, was totally tolerant, ignoring such experimentation even when Uranus was five or six years of age.

Siswi was the first offspring of a wild-born, ex-captive mother at Camp Leakey. She was also the first daughter of an ex-captive to become a mother herself, at least among the orangutans who frequent Camp Leakey. Her mother, Siswoyo, had belonged to the former police chief of the Republic of Indonesia and arrived in camp as a thin, almost gangly, adolescent who had lived in a cage for so long that she could not stretch her limbs and could only walk in a squat . By the time she had given birth to Siswi, she had regained her strength and filled out. Eventually, she became the most dominant of all ex-captive females associated with the camp. Despite her many years in captivity, Siswoyo was a superb mother. No matter how friendly she was with people, she wouldn't let anyone handle Siswi, although she allowed a few people to touch her infant. A visiting scientist once found out just how protective Siswoyo was when he attempted to take a pinprick of blood from infant Siswi's big toe. Siswoyo lunged

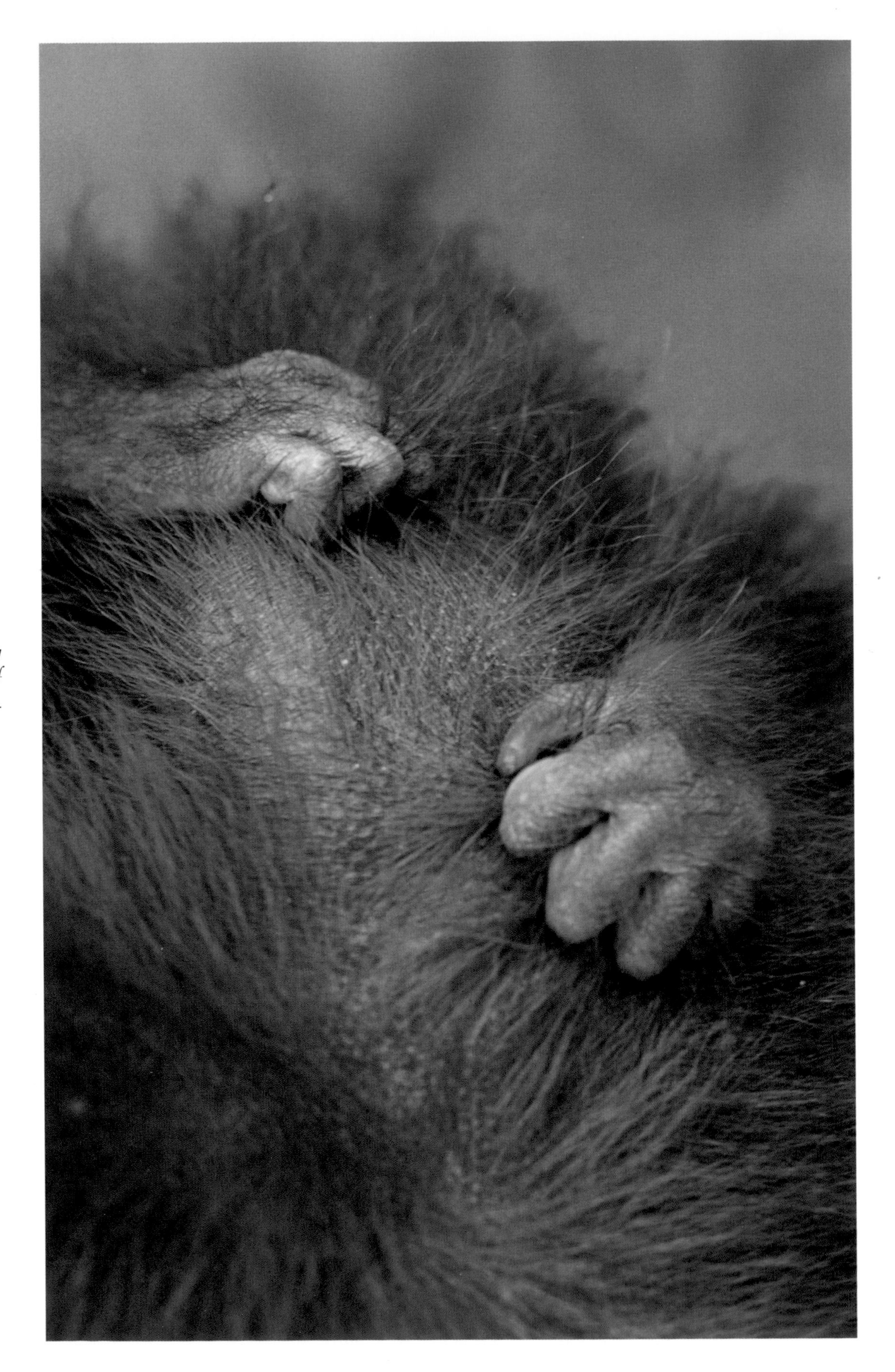

RIGHT AND OPPOSITE: *Until young orangutans reach their fifth year of life, they cling to their mother whenever she moves from tree to tree.*

Following a sudden rain, a suckling infant ignores his mother's surprised reaction. Orangutan youngsters may suckle as long as eight years.

at the scientist, her fangs bared. He reacted quickly enough to escape unharmed. At another time, however, Siswoyo allowed the same scientist to prick her own big toe, not once but twice, for blood samples. It may have helped that she was distracted on that occasion by being given a can of thick chocolate syrup.

Years later, when Siswi gave birth to her first infant, Sampson, she was as protective as her own mother had been. An unfortunate accident involving a park employee cost Sampson his right eye, and Siswi's tender care of her infant son could not have been more concerned and loving had she been a human mother. Immediately after the accident, she would not allow anyone to touch the blind eye, now leaking with optic fluid. Amazingly, she accepted a tube of ointment I offered her and attempted herself to pat the salve gingerly around his eye. By this time I had seen numerous examples of how smart orangutans are, but this took my breath away. The cautious tenderness with which she stroked the ointment on the area around Sampson's eye, the concern and intelligence in her nursing, were the essence of devoted motherhood.

By the age of six or seven, an orangutan juvenile has never been out of his mother's sight. The mother is unswerving in her mothering. But by now juveniles are traveling on their own behind the mother, still suckling occasionally but largely independent for food. The world is still calm. Then, abruptly, the juvenile's world is shattered. The mother begins to deny the juvenile access to the nipple. Weaning begins. The process ends when the adult female becomes pregnant again and gives birth to a new infant. Although the juvenile may travel with his mother for another year and play with and sit behind her, there is no turning back. No matter how gentle or how tender the mother's ministrations are, her concern is focused on the helpless infant who now clings to her night and day. For the juvenile a new chapter begins, a chapter that will lead to a period of intermittent gregarious association with other juveniles and adolescents who also have strayed from their mothers. The chapter ends when the immature orangutans become adults: females taking on dependent offspring, and males developing the cheekpadded faces that identify them as fully grown.

OPPOSITE: *A wild orangutan mother and infant stare calmly out from the foliage.*

The wild mother and infant I followed that day encountered no other orangutans. By midmorning I recognized her as Priscilla. Previously, she had been so high in the canopy I couldn't see her face clearly, only her toes and chin. Her infant was Phil. In the past, an older child of Priscilla, a female adolescent incongruously called Pete, occasionally had come by. But she didn't today. Priscilla spent most of the day foraging in trees with fruits and flowers. Twice she ate young leaves. Phil was tucked under her arm much of the time. It was a typical orangutan day. Infant Phil never once squealed. That night I heard Priscilla cooing gently to her infant in the nest. I knew from watching Unyuk, Kuspati, and Princess what the soft sound meant. At the end of a calm day foraging and feeding, Priscilla was urging her infant to suckle.

With her infant protectively clutched to her body, the adult female called Unyuk at Camp Leakey warily eyes an approaching orangutan.

An orangutan mother's pursed-lip kiss-squeak of annoyance erupts into an open-mouthed shriek as a snoozing infant sleeps undisturbed on the back of her neck. At first, she purses her lips to make a vocalization like a passionate kiss. Her further gesture reveals formidable canine teeth, which, as large as they are, seem modest by comparison with the canines of an adult male orangutan. This female's strong, undecayed teeth show the prominent stains typical of wild orangutan teeth. The infant remains unperturbed through both gestures.

With her large infant sitting docile, an adult female hangs from a small tree. Her extended arm is longer than her head and trunk.

OPPOSITE: *Wild adult female Remaja kiss-squeaks her protests at human observers as she views a wild Bornean bearded pig moving on the ground toward her. Infant Rocky keeps a hand on her mother for reassurance. High up in the trees, the female and her infant are in no danger, but they are on the alert.*

Suspending her body like a hammock, this orangutan mother supports her infant with ease.

The adult female Kuspati, raised at Camp Leakey, embraces her son, Kris. Kuspati's fiery red hair is typical of Bornean orangutans. Her expression spells impatience with Kris's energetic bounce. He is four years of age and is deeply attached, like all four-year-olds, to his mother.

ABOVE: *Upside down like trapeze artists in a circus defying gravity, two juvenile males gleefully continue their play-fight in the trees.*

LEFT: *A juvenile male drops sand on himself, totally lost in play. Juvenile orangutans are extremely playful, but in captivity their natural rambunctiousness and strength translate into destructive behavior. This is among many reasons why orangutans and other large primates should not be kept as pets.*

OPPOSITE: *Framed by the trees, a calm ex-captive mother and infant, accompanied by an older sibling, quietly observe passersby.*

Juvenile males play all out, preparing for the male-to-male competition that will be so crucial in their future. Both exhibit play-faces. The small juvenile male's play-face and throaty chuckle indicate the fun he and his pal are having play-fighting.

Chapter 3

KUSASI: ADULT MALE ORANGUTAN EXTRAORDINAIRE

The next morning before dawn I arrived at Priscilla's nest. Even though I was two degrees south of the equator at sea level, I shivered in the darkness. The temperature had plunged twenty degrees during the night as was typical during the dry season. Up in her night nest Priscilla and infant were cozy as Priscilla had covered herself with two or three leafy branches. An hour or two after dawn, the sun would warm up the morning to the heat of the tropics. In the meantime, I hugged myself, yearning for a warm woolen blanket. As I continued huddling in the dark, I discerned the faint throbs of an adult male's long call in the distance, reverberating in the morning chorus of birds and animals that preceded the dawn. Priscilla and infant did not stir. I wondered who the calling male was. Perhaps Priscilla knew, but if she did, she gave no sign. The nest was silent.

Nothing prepared me for my first encounter with an adult male orangutan in the wild, not pictures in glossy magazines, not even seeing orangutans in a zoo. Adult male orangutans are among the largest of all the great apes, rivaled only by adult gorilla males, which weigh about 400 pounds. Male orangutans encased in the blubber of an easy life in captivity may weigh as much. Ranging between 200 and 300 pounds in the wild, an orangutan male projects a daunting presence, especially if he is suspended directly above the observer.

Kusasi, a subadult male who has not yet begun to grow his adult cheekpads, shows scars and cuts on his face that are the result of tussles with other males.

When an adult male orangutan stands up on his short legs he is only about five feet tall, but his arm span may range between six and eight feet, giving him an apparent stature relative to humans that far exceeds his height. These long, powerful arms enable him to spend hours dangling from tree branches, eating leisurely, resting, or sometimes snoozing. The impact of a male's size is intensified by his face-framing cheekpads. The large, shallow, dish shape of the male's face, encircling his small glittering eyes, brimming with intelligence, gives him a futuristic look, as though he had just stepped off a spaceship. It completes the picture of an alien life force trapped in the treetops. When he is agitated or vocalizing, his usually loose throat pouch inflates, giving the bizarre impression that he has a beach ball stuck under his chin. The effect is unearthly. When I gaze up at adult male orangutans—improbably huge, clad in bulky orange "space suits," tiptoeing along the branches of the upper rain forest canopy—I wonder why people spend so much effort searching for intelligent life in space when there are such intelligent aliens right here on earth.

As often as I have seen adult male orangutans in the forest I still cannot help being impressed by them. No wonder the aboriginal Dayaks and the more urbane Melayu,

In a play-chase, subadult Kusasi chases a juvenile male, who rushes toward his mother in distress.

as well as nineteenth-century Western scientists, thought mature male orangutans were a separate species, distinctly different from adult females and subadult males. At two and sometimes three times the size of an adult female, the adult male's bulk is of a different magnitude.

Adult males are the most solitary of all orangutans (probably contributing to the belief that they were a different species). My quarter century of research has documented that adult males only become social in the company of females interested in mating. And sexually receptive females are rare in the forest since wild female orangutans spend years breast-feeding and rearing their offspring and give birth on average only once in eight years. The scarcity of females ready to breed also helps to explain the male orangutan's extraordinary size and appearance. To father offspring, an adult male orangutan must range over a large area, maintain his rank among other males on the same quest, and defend his prize from other suitors when the rare opportunity

Silhouetted against the darkening sky like Indonesian shadow puppets, Kusasi and another subadult male play in a tree.

Kusasi, on the left, who has now begun to grow his cheekpads, grooms a female as Yayat, a male who is almost a fully-grown adult, views the scene suspiciously.

to mate occurs. Thus, size counts. It is perfectly clear to me that a male's huge bulk, powerful musculature, and forbidding appearance are the marks of the aggression necessary in the extreme sexual competition of the orangutan species, and not, as some scientists have claimed, a relic of a life on the ground during the Pleistocene era.

To mate successfully orangutan males must fight to keep other males away. Orangutans fight exclusively one-on-one, similar to what happens among gorilla male groups, where a single silverback gorilla male usually confronts an individual silverback male trying to take a female from the group. This is unlike wild male chimpanzees, whose related males are bonded into communities. Male chimpanzees devote much of their time, especially in adolescence and in their prime, to forming and maintaining bonds with other males in their community. Groups of males fight groups of males from other communities to prevent encroachment on their fruit trees

A solitary Kusasi stands nearly erect holding onto a tree and looks around before going along a rain forest path. Subadult males become increasingly solitary as they approach full maturity.

and other resources, including females. An adult male chimpanzee's place in his community is not necessarily determined by his size or fighting prowess but by his ability to manipulate others and his success in forging alliances. Survival among adult male chimpanzees depends on highly developed political and social skills.

Because adult male orangutans live primarily alone their social skills are of a different order, though perhaps no less important than among chimpanzees. Orangutan relationships are not with a community or group but with individuals. This is one reason why I grew to like orangutans so much. When orangutans bond, they bond strictly with another orangutan; when they bond with humans, they bond with one person only. Their relationships are extremely individual and extremely personal. In the intense male-to-male competition over individual females, an adult male orangutan can draw only on his own resources. He cannot shift part of his responsibility to his brothers or his allies. The strength of a male orangutan comes from his knowledge, whether conscious or evolutionary, that he is a self-contained unit, totally responsible for himself.

The arm span of human beings roughly equals their height. Here, a grimly alert Kusasi displays an arm span that greatly exceeds his head-to-toe measure. Arm span frequently reaches eight feet in fully grown orangutan males.

Let me tell you about one quite extraordinary male orangutan:

Kusasi, now in his mid-twenties, is at the peak of his power. His almost perfectly symmetrical cheekpads give his face the look of a full moon at dawn. His solid, squat, muscular body weighs at least three hundred pounds. He is strong enough to break a man's neck with one snap of his wrist, or to tear off a person's arm with his teeth, as captive orangutans have occasionally done in their frustration with zoo life. Kusasi's grandeur and presence did not come easily. One of his canine teeth, a male orangutan's chief weapon, has been snapped off at the root. His face and his body under his flowing coat are covered with almost imperceptible scars that testify to the numerous battles he has fought to achieve dominance. When Kusasi strides into Camp Leakey, everyone, whether human or orangutan, pays attention.

Kusasi was born in the wild, but his mother was killed and he was taken captive as a small infant. Fortunately, he was not in captivity for long, because the local representatives of the Indonesian Forestry Department were actively confiscating captive orangutans, with our strong encouragement. It was the late 1970s. The tropical rain forests of Kalimantan were still relatively intact. Small, locally owned timber companies had begun to use bulldozers and chain saws but did not clear-cut vast areas of the forest, as is happening today. The destruction of forest habitat and exposure of wild orangutans to capture in Kalimantan Tengah (Central Indonesian Borneo) were on a relatively small scale.

When Kusasi was brought to Camp Leakey, I was away. Gary Shapiro, then a graduate student working with me, was in charge. He put Kusasi in one of the quarantine cages. Our practice was to quarantine new arrivals for at least two weeks to a month so that we could deworm the newcomers and assess their health. The quarantine cages were simple wooden structures at the back of camp.

Kusasi's first week passed uneventfully. The local assistants, who fed the orangutans in quarantine and monitored their condition, reported that Kusasi was doing well. One day around noon Shapiro decided to check on Kusasi's cage. To his horror,

It has been thought that orangutans are afraid of water, but as this sequence shows, Kusasi may be cautious, but he is not reluctant to enter the river and cross it. Surveying the scene at first, he then enters the river standing upright and wades gingerly toward the deeper part of the water. Having waded up to his armpits in the Sekonyer Kanan River, Kusasi stares across to the other side, contemplating his next move. Eventually, he gets across successfully by using his long arms to catch a bit of vegetation from the other side and then pulling himself across.

he found that the door was open and the cage empty. The ground around the cage bore the fresh hoof marks of a wild pig, and the grass nearby was flattened, as if the pig had dragged something from the cage into the forest beyond. The Bornean bearded pig is an omnivorous creature with a massive snout, partially decorated with a length of short coarse hair. Usually light in color, the pig can stand more than three feet tall at the shoulder. We knew that some years earlier a wild pig had killed and eaten a juvenile orangutan in the forest nearby. I had seen a wild pig dragging the carcass of an aged wild orangutan in the forest: dragging dismembers the body so the pig can eat it more easily. The logical conclusion was that Kusasi had met a similar fate.

After I returned to camp and Gary told me what happened I made an official report to the Indonesian government that Kusasi was dead. Time passed. Then, one day an assistant ran up to my hut out of breath. Eyes wide, he told me that a wild infant orangutan was clinging to the wire mesh that covered the windows of the dining hall. "Where is the infant's mother?" I asked. The assistant said there was no mother. "Impossible," I replied. If a wild infant orangutan was in camp, its mother had to be nearby. I hurried back to the dining hall with the assistant. I saw no sign of a female, but sure enough, a large infant, scrawny as a plucked chicken, was clinging tightly to the mesh over the window. I stood there perplexed. I had never seen the infant before; I thought that this could not be an ex-captive. But if the infant had appeared from the forest, where was the mother? Orangutan mothers never let their infants wander off this way. There had to be a mother.

A few minutes later, Shapiro strode by intent on some task. He barely glanced at the young orangutan clinging to the mesh because at that time ex-captive orangutans occasionally loitered around the dining hall, waiting for an opportunity to beg or steal food. Suddenly Gary stopped, turned around, and stared. "Oh my gosh," he exclaimed, "it's Kusasi!" It took a moment for me to register what he had said. More than a year had passed since Kusasi had disappeared. "Are you sure?" I asked, for I had never seen Kusasi. "I thought you told me he was dead." "He was," Gary fumbled. "I mean, I thought he was, but I never saw the body. No doubt about it, that's Kusasi."

We were flabbergasted. Kusasi had spent almost a year and a half on his own in the forest. He was now probably four and a half, perhaps five years old, the age at which wild orangutans no longer cling to their mothers but follow them closely as they move from tree to tree. One of the first wild-born, ex-captives I had released at Camp Leakey, Akmad, disappeared at about the same age. But she had been in camp for half a year or so before she returned to the forest. She was a small juvenile. Kusasi had been a large infant of about three when he vanished. Perhaps Kusasi attached himself to a wild adult or adolescent female. I have seen female orangutans treat younger orangutans almost like their own offspring. Adult females with infants of their own occasionally allow an unrelated ex-captive infant or juvenile to tag along, sharing some of their food and perhaps even their nest with the interloper. Even a stepchild association with an adult female orangutan and the scraps of food that she brings may help a motherless orphan cling to life. We'll never know how Kusasi survived—or, for that matter, why he reappeared in camp. It may be that he developed skills precociously after the death of his mother and became independent at an exceptionally early age. Presumably, he was released from his quarantine cage by an ex-captive orangutan when no assistants were present in camp.

Even after his return to camp, Kusasi remained independent. In those early days I personally mothered many of the wild-born ex-captives who came to Camp Leakey. Some glued themselves to my body, as they would to an orangutan mother, night and day. Other juvenile orangutans regularly sought solace and comfort in my arms. I never mothered Kusasi in the same way. He didn't like to be carried nor did he care very much for human touch. His attitude toward other people ranged from indifferent to hostile. Kusasi was primarily interested in food. He came and went, spending longer and longer periods in the forest, but always returning.

As an adolescent, Kusasi formed very strong friendships—especially with another male ex-captive in camp named Pola and with some of the adolescent females in camp. Compared to Kusasi, Pola had led a sheltered youth. One of Gary Shapiro's "students," he had learned sign language as a juvenile. Although not as proficient as Gary's star pupil, Princess, he acquired a repertoire of about a dozen signs, which he

occasionally combined to form short phrases. Pola was calm, even tempered, good-natured, and totally relaxed with humans. Kusasi was the opposite: belligerent, jumpy, quick-tempered, and standoffish. Pola trusted people. In contrast, Kusasi never allowed anyone near him but me. Unpredictably, he would charge at assistants, and he invariably greeted unfamiliar visitors to camp with arm-flailing, aggressive displays. Occasionally, Kusasi stalked visitors, apparently enjoying appearing out of nowhere, which sent everyone screaming in all directions, although he never harmed anyone. Once or twice he also grabbed people by the ankles and toppled them, but did not follow up with any kind of attack. After holding them, he merely let them go. For some reason, though, he particularly disliked a police sergeant who usually came to Camp Leakey in uniform. Kusasi's charges on those occasions caused the police officer to shed all dignity and run.

As subadult males, Pola and Kusasi both avoided the wild adult male cheekpad-

ders who sometimes passed by camp. But they seemed almost obsessed with one another, seeking each other's company and apparently never tiring of being together. Their play-fighting was quite rough but it was clearly play. At such times both showed the characteristic orangutan play-face, complete with broad grins and bared teeth (perhaps the origin of the human smile); at times I heard deep, throaty chuckles that almost sounded like human laughter. But there was an edge—a seriousness—to their play that I didn't see in the play of female juveniles and adolescents. I got the impression that play-fighting would determine Pola's and Kusasi's destinies.

During this time Kusasi and Pola shared matings with Camp Leakey's wild-born ex-captive females or their adolescent offspring. Each had his own favorites, with whom he mated almost exclusively.

As the two males got older, their forays to the forest increased. But they seemed to live different lives in the wild. Pola almost always returned from the forest without

a scratch. Kusasi, on the other hand, came back so cut up he looked as though he was regularly involved in gang fights. Almost every part of Kusasi's body was bitten, ripped, or slashed at some point. There were times when he limped back to camp so torn and shredded that I thought he would die. But Kusasi always allowed me to medicate him, and he always returned to good health. They say what doesn't kill you makes you stronger; thus it was with Kusasi.

Then, at some point, the equation began to change. A new subadult male, Yayat, was released at Camp Leakey. Kusasi became as obsessed with Yayat as he had been with Pola. Kusasi and Yayat play-fought ceaselessly. Pola began to disengage from the fray, spending more time in the forest. Then, in a matter of weeks, Pola sprouted cheek-pads. He did not increase in size nor did his temperament and character change, but I felt certain that he would reign supreme. Though Kusasi clearly was larger and more

Other orangutans approach the water and react to it in different ways, displaying caution but not fear.

aggressive, he was not overwhelming. Pola's serene sense of purpose and calm determination seemed to assure his dominance. It was not by coincidence at this time that the resident adult male who had dominated Camp Leakey quietly disappeared.

Suddenly, however, new and larger adult males materialized, many of whom were ex-captive graduates of Camp Leakey. Their fighting skills honed in the forest, they soon drove Pola away. Pola reappeared in Camp Leakey once or twice but then vanished into the forest. Now, the rivalry between Kusasi and Yayat, both still subadults, escalated. Not only did they spend hours play-fighting, they also engaged in bouts of competitive grooming. Orangutan grooming is usually brief and gentle, but in the case of Kusasi or Yayat, whichever was dominant at a given moment forced the other to submit to long bouts of grooming. It was as if one had said to the other, "You're defenseless. I'm going to groom you whether you want it or not. Belly up!"

On the bridge at Camp Leakey, which he visits periodically, Adik stands upright on his hind legs totally at ease. A brachiator's anatomy preadapts orangutans and other apes to occasional bipedal posture. (Photograph by Michael Charters)

OPPOSITE: *Broad cheekpads distinguish the magnificent adult male Adik, who sits on the forest floor gazing pensively. Old wounds from previous combats with adult males are apparent around his mouth.*

Before long, Yayat became a mature cheekpadder. Unlike Pola, he became larger and more muscular in the process. Although Yayat was obviously stronger and more mature than Kusasi, the two continued to play-fight. Kusasi seemed somewhat afraid of him but would not give up. He refused to leave the forest around Camp Leakey.

Kusasi also grew. Long, lean, and lanky, he became one of the largest subadult males I have ever seen in the wild. Still jumpy and irritable, his temperament did not improve. Once, when I touched him, he slapped my hand. He did not charge or bite, though. He was simply letting me know that he was a big boy now. Around camp, Kusasi continued frequently mating with the wild-born, ex-captive females or their adolescent daughters.

When I went to North America one summer, I left Kusasi as a subadult male. When I returned a few months later he had developed the full cheekpads of a mature male. Most amazingly, he had filled out, almost as if he had taken steroids. His rock-solid body, with rippling muscles on his back, now dwarfed Yayat.

A series of battles over the next year established Kusasi's dominance beyond doubt. Always exceptionally calm, Yayat remained so in defeat. Although he spent most of his time in the forest, he did not vanish. He sometimes visited the camp, but when Kusasi approached, Yayat faded into the forest. What amazed me was that in their battles Kusasi did not seem to bite Yayat but rather pummeled him hard. Was the long play-fighting friendship of subadulthood responsible for Yayat not leaving, even though Kusasi clearly was king? Was Kusasi's relatively benign aggressive behavior shaped by his long association with Yayat? I don't know, but I believe that it is possible that somehow the relationship that had been established in subadulthood had continued into adulthood.

Kusasi's character changed in adulthood. With dominance achieved, he mellowed. Now he comes and just lies outside my house at Camp Leakey, peacefully gazing up at the sky and the crowns of the trees. The sneak attacks and jerky indecision of his youth have disappeared. Graceful and smooth, Kusasi materializes out of nowhere and makes himself at home. Then, in the blink of an eye, he seems to vaporize. Self-assured, dignified, almost regal now, Kusasi sometimes reaches out to touch people very gently. More than curiosity, the gesture is a statement of equality and acceptance.

Kusasi's life recapitulates that of a wild male orangutan. I've seen rough and tum-

In response to a tense moment, the adult male Pola yawns, showing his formidable canines, as he confronts Kusasi.

OPPOSITE: Now displaying the cheekpads of a full adult, Kusasi reveals the toll that competition with other males has taken: two of his canines, one upper and one lower, are broken. Female orangutans virtually never display broken canines. Here Kusasi is yawning in a tense situation.

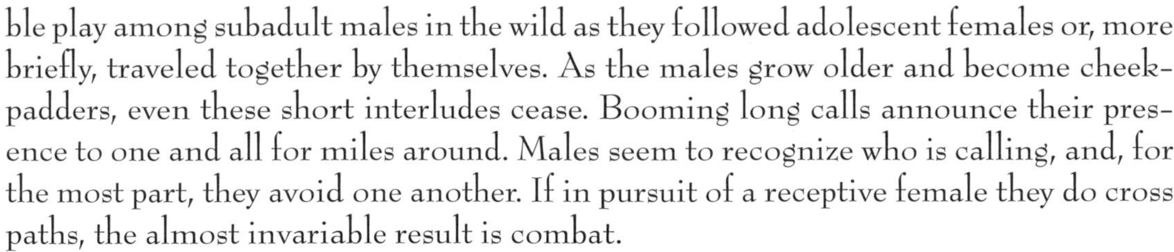

ble play among subadult males in the wild as they followed adolescent females or, more briefly, traveled together by themselves. As the males grow older and become cheekpadders, even these short interludes cease. Booming long calls announce their presence to one and all for miles around. Males seem to recognize who is calling, and, for the most part, they avoid one another. If in pursuit of a receptive female they do cross paths, the almost invariable result is combat.

Looking as gigantic as King Kong in comparison to females, Kusasi approaches Siswi, an adult female with whom he has been friendly since his youth, and her son, Sampson. Adult male orangutans seldom pay any attention to their former mates, but here Kusasi seems to reach out to Sampson, who may be his own son.

These combats are rare, however. In a quarter of a century I have only personally witnessed a dozen or so. Sometimes the fights last only a few minutes, with one of the males turning tail and fleeing before any damage is done. At other times, especially when the males have a history with one another, the confrontation may last off and on for hours. That orangutan males invariably fight one-on-one in part explains their massive size, but there may be an additional explanation as well. Opportunities to mate are rare. When a male does enter a consortship with a receptive female, he shows little interest in food but concentrates on guarding and monitoring the female. Once the consortship is over, the male is alone again. The male's massive size not only allows him to win combats with smaller, less powerful males but gives him reserves of body fat and weight so that he requires less food during critical times such as consortship. When the adult male orangutan is alone, he concentrates on foraging and eating to build up his reserves. His days are spent preparing physically and, I suspect, mentally for the next opportunity to mate and the next encounter with a competitor.

As I followed Priscilla that day, I heard a twig snap in the distance. It was now late morning. In her fourth food tree of the day, Priscilla didn't even turn her head but continued feeding on succulent, almost translucent young pink leaves in a low tree. I moved quietly through the forest, searching for the cause of the sound, hoping it was the male who had called just before dawn. I spotted a troop of leaf-eating monkeys fifty yards away, their thick red coats almost glowing in the shadowy understory. Then a twig snapped again. I gazed up. A fully cheekpadded male was sitting midway up a tree, eating clusters of berry-sized fruit plucked from a thick vine. If he noticed me, a small, puny primate of no significance to his life, he gave no sign. Someday, I thought to myself, if all goes well, Priscilla's infant son Phil would achieve the same monumental majesty. I was momentarily saddened, because Phil as an adult would likely be miles away, and when he returned as a cheekpadder, he might be unrecognizable. I gave a note of thanks to Kusasi, who had returned and stayed and heightened our understanding of male orangutan life.

OPPOSITE: *It is no coincidence that Kusasi's head-scratching gesture, bemused smile, and direct gaze strongly echos similar body language in humans.*

Chapter 4

BORNEO'S GREAT FOREST

I think that I shall never see
a poem as lovely as a tree.
Poems are made by fools like me,
But only God can make a tree.
Joyce Kilmer

Since its discovery by Europeans at the end of the Middle Ages, the island of Borneo has haunted the Western imagination. The establishment of a white raj, or kingdom, by a Victorian British adventurer heightened the romantic allure of the island. Teeming with exotic creatures and covered with vast, seemingly endless, fecund forests, Borneo constituted the seeming antithesis of straitlaced Victorian England. Even compared with America and Canada, Borneo was relatively untouched. As the Pony Express galloped across the vast North American plains to the Pacific Coast in the middle of the nineteenth century, followed soon by even speedier iron horses, Borneo remained untamed, its half-naked Dayak hunters reputedly ready at any moment to savor human flesh. Margaret Mead, the famous anthropologist, remarked that her parents only used one racial slur. When she hadn't combed her hair to their satisfaction, they told her she looked like "the wild man of Borneo." Ironically, the wild man of Borneo was probably the orangutan.

The orangutan's largest homeland lies in Kalimantan, the Indonesian part of Borneo that takes up about three quarters of the island. Borneo is the third largest island in the world after Greenland and New Guinea. Unlike most of Indonesia, Borneo lies outside the volcanic belt known as the Ring of Fire. Although Indonesia has seen several major volcanic eruptions in the last century, Borneo itself has remained geologically quiet for at least fifty thousand years. This has resulted, however, in Borneo having relatively infertile soils. The neighboring island of Java, with its rich volcanic soils, has the highest population density on the planet; Borneo has one of the lowest. Kalimantan accounts for almost 30 percent of the landmass of Indonesia but less than 5 percent of its population

The majority of Kalimantan's people live in coastal towns and cities. Two of the most populous and oldest cities, Banjarmasin and Pontianak, appear as counterweights, with Banjarmasin on the southeast and Pontianak almost directly opposite on the northwest coast. In the interior, Dayak villages hug the riverbanks, isolated by the extremely steep hills and valleys that slice through the terrain. The course of Kalimantan's navigable rivers can be charted by linking lines of villages from the seacoast north into the interior. The spine of the island, the Schwaner mountain range, is the

A small boat carrying Camp Leakey research assistants penetrates the early morning haze, thickened here by the smoke of the 1997 fires.

The black water of the Sekonyer Kanan River reflects like a mirror, as it slices its way through prime orangutan habitat, peat swamp forest.

source of Borneo's mightiest and longest rivers. Yet much of Kalimantan, especially the coastal areas and the many deltas where these rivers empty into the sea, is low and swampy. From the air, the lowland forests of Tanjung Puting National Park resemble an endless ocean of green stretching to the distant horizon.

The origin of the name Kalimantan remains unknown. Some claim that the word means "river of diamonds"; others suggest that it means "many rivers." The word Borneo is acknowledged to be a European corruption of the word for Brunei, the long-lived Melayu sultanate that once ruled over much of Borneo's northern coastline and thus controlled access to trade in the interior. Despite Borneo's distance from the main centers of civilization, trade with the ancient kingdoms and empires of China and India flourished off and on for thousands of years. But the ancient kingdoms of Borneo, such as Brunei, did not penetrate very far into the interior.

Until recently about 90 percent of Kalimantan consisted of primary tropical rain forest, the world's second-largest continuous expanse of tropical rain forest after the Amazon Basin. The great nineteenth-century naturalist Alfred Russel Wallace described the tropical lowland evergreen rain forest of Borneo as a cathedral forest. Some of the tallest trees have almost perfectly cylindrical trunks that soar almost one hundred feet into the sky before branching out. Below, this forest is dark and dank, and the ground vegetation is sparse. The somber parkland atmosphere remains totally at odds with the popular notion of a jungle, a tangled mass of impenetrable vegetation overrun with ferocious animals, brilliantly colored birds, and poisonous snakes. In reality, few of these rain forest animals are seen, for they are masters of concealment and camouflage, even when they move. The most ferocious organisms are usually also among the smallest. It is not tigers or pythons that are to be feared by humans but ticks, leeches, and mites on one hand and bacteria and viruses on the other.

Tropical rain forests differ from most temperate forests in a variety of ways. One of the most basic differences may not be obvious to the casual observer. Ninety percent of the organic matter found on the tropical rain forest floor is recycled every six months, a process that in other forests takes three years. A tree may totally disappear within several years after dropping to the forest floor. This energized recycling of nutrients is so complete that no deep layer of humus builds up and the soil tends to be barren, with a relative paucity of leaf litter.

Tropical rain forests have been called a counterfeit paradise, in part because the lush vegetation belies the poverty of the soils below. More than 60 percent of the humid tropics have infertile soils. If the trees and other vegetation are cut down or burned, the ground eventually turns to sand. Although many more kinds of fruit are found in tropical rain forests than in temperate areas, relatively little fruit is usually available at any one time. Yet tropical rain forests contain an amazing diversity of plant and animal life. Covering only about 7 percent of the earth's surface, tropical rain forests are home to at least half the world's known animal species, including 75 percent or more of all insects. Borneo is five times the size of England and Wales, but it has 5,000 known species of trees compared to only 34 species of trees native to England. Seven hundred and eighty species of trees were counted in a single 25-acre plot of forest in Borneo, a measure of the diversity of tropical rain forest habitat.

Hanging casually from a vine, a juvenile orangutan examines an edible morsel found in the lower tree canopy.

OPPOSITE: An adult female examines a kayu putat flower before eating part of it.

The richness of species is complemented by the complexity of the physical structure of the forest. There, towering trees with buttressed trunks support masses of lianas and creepers as well as epiphytic orchids and bromeliads. The structure of this forest is usually described in terms of its layers. In dissecting the layers of the tropical rain forest like a piece of cake from top to bottom, the uppermost is the emergent layer. German naturalist Alexander von Humboldt called this layer "a forest above a forest."

Next is the canopy, typically a blanket of the contiguous crowns of large trees that block out most of the sunlight. In some cathedral forests, the canopy is so dense that only 5 percent of the sunlight reaches the lower layers of the forest. Most of the trees in the canopy will never grow large enough to reach the emergent layer. The understory consists of a variety of shade tolerant trees, including some pole trees that are the younger individuals of species that will someday reach the upper layers of the forest. Finally, the herbaceous plants, tree seedlings, and saplings on the forest floor make up the ground layer. Where the forest opens up, for example on riverbanks, this layer can be exceedingly rich and lush.

Most of the towering trees of the emergent and canopy layers in the tropical rain forest do not have deep penetrating tap roots; rather, their trunks are supported by plank and flying buttresses. In the peat swamp forests along Kalimantan's coasts, stilt roots, aerial roots, and knee roots are common adaptations to the seasonal flooding that often lasts six months or more.

Lianas and epiphytes are common throughout the tropical rain forests. Most vine species or lianas are found in tropical forests. Lianas sprout on the forest floor and climb upward. Lacking woody trunks, they grow like wreaths around young trees, hoisting themselves toward the sunlight. The most common lianas in Borneo are the rattans (climbing palms). Some large lianas look like thick, twisted ropes and easily support the weight of even the largest adult male orangutans. The weight of lianas themselves may cause tree limbs to break and fall.

There are at least thirty thousand known species of epiphytes in the world's tropical rain forests. Epiphytes—including mosses, orchids, ferns, lichen, bromeliads, and even cacti—need wind to spread their pollen, spores or seeds. Epiphytes virtually always grow on other plants, usually trees. Among the most common in Borneo are staghorn and birdnest ferns, which often plant themselves against a tree trunk or in the fork of a branch. Some epiphytes have dangling roots that absorb water from the moist air and the frequent rainfall. Most epiphytes do not harm the host trees. The main exception is the strangling fig, which starts out as an epiphyte, sending its roots to the forest floor. These roots increase in number and size and eventually encase the host tree. The strangling fig shades the crown of the host tree and finally causes the host to die. In many parts of Borneo and northern Sumatra (though not in the area around Camp Leakey) the fruit of the strangling fig is important in the orangutan diet. Sometimes half a dozen or more wild orangutans will congregate in a large fig tree, enjoying the plentiful fruit.

People brought up in the temperate regions of Europe, Asia, and the Americas view the passing of the seasons—spring flowers developing into summer fruit and vegetables, then dropping seedpods in fall and becoming dormant in winter—as an integral part of life. The tropics have only two seasons, wet and dry. But these seasons do not precipitate flowering and fruiting. In the tropics different plant species have different cycles, although sometimes, especially after dry spells, more species flower than at other times, and some species fruit annually, but not always in the same month.

In Borneo, as in much of Southeast Asia, the most common large trees in tropical lowland and hill forests are members of the family Dipterocarpaceae. This family includes many commercially important hardwood timber species, some known locally as meranti. Although they are found elsewhere in Asia, Borneo has more species of dipterocarps, and more endemic species, than any other place in the world. The emergent layers of Borneo's forests contain a high percentage of dipterocarps as well as legumes (the family that includes peas and beans). The most common type of lowland forest in Borneo is a cathedral forest dominated by dipterocarps. At Tanjung Puting, where the soil is especially poor, the forest canopy is lower and the trees are smaller. Without a closed canopy, more light reaches the ground, and the forest is more open. Tanjung Puting primarily consists of tropical heath forest, also locally known as kerangas, and peat swamp forest. In these forests the layers of the primary rain forest intermingle. Emergents are rare. The canopy and the understory sometimes cannot be distinguished. The relatively open canopy encourages dense, tangled vegetation so that movement at ground level is difficult and visibility is poor. Insectivorous plants are abundant, including the pitcher plant, Nepenthes.

The most extensive heath forests in the world are found in the upper reaches of the Rio Negro, the large black-water tributary of the Amazon River. Black-water rivers frequently are found in association with tropical heath and peat swamp forests. So it is in my study area. The Sekonyer River, which drains the northwest part of Tanjung Puting National Park, is a black water river, so called because in a certain light the water looks black and almost opaque. But when seen in a glass, the water is tea colored. The water itself is clear but the dissolved organic materials and acids give it color. If solidified, the water would resemble dark Baltic amber.

In parts of coastal Southeast Asia, conditions that have existed since the Pleistocene era promote the formation of peat swamps. Due to unusual salty and waterlogged conditions, the leaf litter fails to decay, resulting in the relatively continuous formation of peat. Tucked behind the mangroves, these coastal peat swamps look as if they were straight from primeval times, with massive epiphytes, tree roots rising out of black water, tree limbs draped with lianas, and the haunting calls of unseen birds.

Orangutans are, truly, "people of the forest." No other great ape spends as much time in the trees. The African apes—chimpanzees, their gracile cousins the bonobos, and gorillas—all have evolved adaptations to terrestrial locomotion: they knuckle-walk on the ground. In a sense, they have their cake and eat it, too: they are at ease both on the ground and in the trees. Orangutans are almost exclusively arboreal. Four-handed, with hip joints as flexible as their shoulders and enormously powerful arms, they may swing hand over hand (called brachiation), or hang from one hand, from one hand and one foot, by both feet, or any other combination of hands and feet while they feed. A wild orangutan female can spend months in the trees, drinking the

Bedecked with a branch, a juvenile orangutan momentarily stops his climb into the canopy, where he will build a play nest in the trees. He will continue to spend the nights in nests built by his mother until a new sibling comes along, and he is forced to fend for himself. Orangutans are dependent on the forest not just for food but also for their shelter at night.

A juvenile male orangutan constructs a night nest in a palm tree at the edge of Camp Leakey. (Photograph by Rod Brindamour)

water collected in leaves or tree hollows, without once descending to the ground. Orangutans depend on the trees and vines of the tropical rain forest for their food. More than 95 percent of orangutan food comes from the canopy. More than 60 percent of it, if measured by the time expended in foraging for it, consists of ripe or almost ripe fruit from mature trees.

Vernon Reynolds, the British primatologist who made one of the earliest studies of chimpanzees, commented on "The strange lackadaisical way of life, here today, gone tomorrow . . . wandering and roaming through the jungles of Borneo" that seems to characterize wild orangutan life and seems so different from that of the chimpanzees. Yet the leisurely pace of the orangutan's life in the treetops disguises the difficulty of finding food in the counterfeit paradise of the tropical rain forest. In tropical rain forests, the availability of food fluctuates unpredictably. The variety in orangutan diet is remarkable not only among primates but also among all animals. In modern Western cities, human beings can usually choose among about thirty different fresh fruits or berries at the market and about the same number of different fresh vegetables. The orangutans in my study area select from about four hundred food types. Using an intelligence that is among the highest of any animal on the planet, orangutans clearly remember and recognize each food type.

The orangutan diet makes up a complex mix of fruit, nuts, young leaves, bark, sap, insects, shoots, stems, honey, fungi, and even the occasional spider, and it varies markedly from month to month according to the amount of fruit in the forest. If fruit is available, however, orangutans mainly choose it, with the periodic addition of tender young leaves. But fruit, the orangutan's dietary staple, is not always readily available in the tropical rain forest, and it is irregularly distributed not only in space but also in time. Fruit is usually scarce in the tropical rain forest, although it may sometimes be seasonally abundant. As is typical of tropical rain forests, the flowering and fruiting patterns of individual species within the rain forests do not always follow predictable yearly cycles in the manner of temperate forest species. Those tree species that fruit in a given year vary greatly in the amount of fruit produced, even among trees of similar girth. And since the forest consists of a complex mixture of species, even when one species is fruiting, the individual trees of the species may be widely spread throughout the forest.

Fruit scarcity and irregular distribution are such prevalent conditions in the orangutan habitat that they may have led to the development of their high intelligence. It has been shown that among primates, species who eat fruit have larger brains in relation to their body weights than leaf-eating species. Leaves are more uniformly distributed than fruit so leaf-eaters are not under intense pressure to remember the locations and patterns of appearance of leaves. Thus, leaf-eaters can depend on regular patterns of movement through their habitats to maximize their probability of encountering leaves. Orangutans at Tanjung Puting don't operate in this way. They are too slow moving to randomly wander or to use routinized patterns to find the widely scattered fruit trees. They must remember and monitor the locations and productivity of fruit sources, and this need was a major selective mechanism influencing orangutan, and possibly all great ape, intelligence.

How do orangutans find the fruit trees? In Tanjung Puting orangutans navigate through relatively flat, featureless, uniform terrain. Altitude is barely above sea level. It is clear that orangutans in this habitat use arboreal pathways in the canopy in novel combinations to find the ever-changing mix of fruiting trees that constitute their food sources. But elsewhere, since orangutans do not move about the forest randomly, topographical features must direct their movements. At some mountainous locations, such as the rugged site in Sabah studied by John MacKinnon, who pioneered orangutan studies in North Borneo and Sumatra, steep slopes and deep valleys may direct orangutan movement in nonrandom patterns. To the casual observer, these seemingly premeditated movements may mimic the orderings of a high intelligence.

Orangutans seem to be better at remembering and locating specific trees and at navigating the forest than scientists equipped with compasses, maps, and cut trails. John MacKinnon pointed out in the late 1970s that orangutans must "have a good sense of direction, distance, and travel time." That this is true is indicated by the way wild orangutans abruptly veer off from a previous route and travel more rapidly than usual just before reaching the fruiting tree that was their presumed goal.

As for the differences in food-getting and travel between orangutan males and females, the basic rule is that males come and go; females stay. Orangutan males also seem to go longer distances when they travel, up to three miles per day. Orangutan females, on the other hand, rarely range over a mile during a day's travel. And the speed at which the sexes travel differs: male speed is enhanced by the fact that they occasionally come down to the forest floor and walk on all fours. Some adult males are so comfortable on the ground that they sometimes make day nests and rest on the forest floor. At other times, they are likely to leave their night nests in the trees shortly after dawn, descend to the ground and walk to their first food source, and climb up the trunk to reach the succulent fruit hanging from the tree rather than making their way in the canopy.

Orangutans are ripe fruit eaters. Some of the fruits they consume are the forest versions of cultivated fruits eaten by local people. Orangutan diet includes wild durians, mangosteens, banitan nuts, mangoes, merang, belale, jackfruit, snakefruit and wild rambutans. The rambutans are a hairy-skinned type of lichee. In the past some of the wild fruit eaten by orangutans would be gathered by forest-living Dayaks and sold in the regional market place. Now with the ready availability of commercially grown temperate fruit from around the world in the stores and marketplaces of even small towns in Kalimantan, the hand-gathering of wild fruit for sale has disappeared except in the most isolated areas.

Many tropical rain forest trees, especially those in the canopy, show a staggered pattern of fruiting. Thus, a few fruits ripen daily on the tree for a period that may extend up to two months. The pattern of orangutan foraging is to exploit such a tree repeatedly while it is fruiting. Every few days orangutans return to the tree to savor the fruits that have ripened in their absence. Orangutans may be keeping track of dozens of fruiting trees of the same species in their foraging rounds over an area of several square miles. As trees of this species drop the last of their fruit, the orangutans switch to other tree and vine species that are just coming into fruit. At any one time orangutans may be monitoring ten to twenty different species that are scattered over the forest and are at different phases of fruiting. Orangutans plan efficient routes to incorporate as many fruiting trees as they can in their foraging. It seems perfectly clear that somewhere in human ancestry was a creature much like the orangutan with a brain that solved similar problems of finding and processing food.

*One of the stalwarts of river edge vegetation, a pulai tree (*Alstonia sp*). subsists even with wet feet. Trunk and leaves are reflected in the black waters of the Sekonyer Kanan River.*

Finding fruit or other food is, of course, only the first step in consuming it. Once found, foods may require peeling or opening before the edible parts can be had. This may entail simply plucking a fruit with an edible skin from a branch, but other food processing may require the removal of fruit flesh, skins, or seeds from a tough seed or

A bridal veil mushroom, which may reach six inches or more in size, blooms for a few days before the veil rots.

husk, or the separation of inedible or toxic parts of a fruit before eating. Durians may weigh up to one kilogram (2.2 pounds), and they have very thick skins. Unless they are extremely ripe they may be difficult for a human to open without a machete or other sharp tool. Hard-shelled nuts such as banitan (*Mezzettia* sp.) are so difficult to crack that juvenile orangutans frequently can't do it. Sometimes it takes an adult orangutan fifteen minutes to crack open a nut—testimony not only to the orangutan's strength but also to its intelligence. And all this work may be done for the tiny bit of meat inside the nut.

I'll never forget the first time that I saw a female orangutan gingerly picking the fruits of a wild relative of jackfruit, carefully lining the fruits upside down on a large branch (as though they were pies on a shelf), and then waiting for the gooey white latex to flow from the wound in the fruit where she had bitten or pulled off the thick, tough stem. Only then did she reach over and start biting the thick skin of the fruit. She avoided the gooey latex, which would have gummed up her mouth, hands, and hair. Orangutans also use branches as anvils and other tools in the processing of fruit. I have seen orangutans, for instance, processing burrs by rubbing handfuls of fruit on a branch until all the burrs were removed. When I tried to imitate this technique with my bare hands I screamed out in pain.

What does all this information about food getting and eating tell us about orangutan

A small pythonlike snake weaves its way through the underbrush near the river's edge.

intelligence? Before the publication of Charles Darwin's book *On the Origin of Species by Means of Natural Selection* in 1859, many learned people accepted creationism, not just because they believed in the literal truth of the Bible but also because creatures appeared so perfectly adapted to their habitats that they thought only a divine hand could have been responsible for such perfection of form and function. Orangutans, shaggy, slow, even-tempered, and apparently roaming somewhat idly through the forest, almost seemed like a violation of this perfection. A British writer of later date than Darwin, Terry Pratchett, described them as a heap of semideflated inner tubes that one passes in one's eagerness to get to the real apes, the more gregarious chim-

The local Dayaks avoid this species of toad assiduously, because they believe that its skin is poisonous to the touch.

panzees and gorillas. He further noted that orangutan faces showed all the expression of a "surprised coconut." Yet through millions of years of evolution, approximately ten to fifteen million after our human ancestors separated from theirs, orangutans have adapted so well to their environments that their lives seem effortless. But, as we have seen, living in the tropical rain forest is difficult. Human beings became successful in that environment only with the introduction of horticulture. With the simple technology of slash-and-burn techniques to make the most of infertile soils and a deep understanding of the environment, relatively small numbers of *Homo sapiens* have been able to live in tropical environments. But many recent attempts to exploit these habitats with more recent technology have foundered. Orangutans make it look easy. Orangutans have survived in the forest for a very long time. Take the orangutan out of the forest and the comic figure of the inner tubes replaces the magnificent, mostly solitary creature. That creature forages with success in the forest canopy by using a brain as large as any animal on earth except humans, and posssibly dolphins, and one that is, in the end, extremely similar to our own.

A Melayu worker bends to cut through the heavy vegetation that seasonally blocks the Sekonyer River. The butterflies encircling him are sucking the sweat off the man's back. Forest workers and park rangers work hard to protect the conservation areas from destruction, but the number of rangers is small and the inroads on protected forest areas are constant.

OPPOSITE: A crested fireback pheasant, known locally as a chicken of the forest and valued by the Dayaks for its tasty meat, struts across a small opening in the forest.

Sharing the canopy with orangutans are other primates, including the proboscis monkey, one of which is seen here leaping for a branch with her infant clinging to her chest. The long tails of the monkeys are used only for balance. They are not prehensile and are, therefore, not suitable for grasping branches as are those of South American monkeys.

All manner of vegetation forms the diet of orangutans—leaves, fruits, bark, and flowers—and the search for food is constant.

Standing erect, an adult female orangutan samples a trailside delicacy. Orangutans will customarily find food in the canopy, but when they are on the ground edible vegetation may attract them.

OPPOSITE: An adolescent male named Tom relishes the inner shoots of a bakong plant near the river's edge. Bakong, members of the lily family, are the main culprits in clogging the Sekonyer River. Their roots break free and the plants float downstream, piling up at narrow points in the river.

Chapter 5

SAVING EDEN

The creation of biodiversity came slow and hard: three billion years of evolution to start the profusion of animals that occupy the seas, another 350 million years to assemble the rain forests in which half or more of the species on earth now live.

E. O. Wilson

A wild orangutan moves across open-ash covered ground adjacent to a burned forest. During the severe drought of 1997–98 caused by the El Niño weather phenomenon, fires devastated tropical rain forests in Kalimantan (Indonesian Borneo), destroying prime orangutan habitats. Approximately three million hectares (more than seven million acres) of tropical rain forest in Indonesia were burned.

The journalist H. L. Mencken once observed that there is a simple solution to almost every complex problem—and the simple solution is almost always wrong. For the complex problem of orangutan conservation in Borneo and northern Sumatra, the simple solution has been the return and rehabilitation of wild-caught captive orangutans to the forest. In this case, the simple solution is not wrong. It is simply not enough.

Over the centuries the public's image of the orangutan has swung from subhuman monster to noble savage. In recent years orangutans have emerged from the shadows as charismatic megafauna; as speechless but eloquent ambassadors for the disappearing tropical rain forest; and as the forgotten relative we left behind in Eden. Thanks to books, magazines, and popular nature shows on television, orangutans have earned the public's goodwill. But goodwill is not enough. As the twentieth century draws to a close, the orangutan has simply become another victim of global economic forces. Like a high-speed train with no one at the controls, the world's economy speeds toward the millennium, overwhelming everything in its path, destroying habitats, and causing the extinction of animals, plants, and insects. Apart from the physical losses to the future of humanity, the ethical and moral degradation of this destruction is incalculable

The orangutan was added to the U.S. Endangered Species list in 1970. Even at that time, some experts calculated that the number of wild orangutans in Borneo and Sumatra was twenty thousand individuals or fewer. Orangutans are slow in breeding. On average, as we have seen, an orangutan female gives birth at eight-year intervals. Under the best of conditions, she will bear only a few offspring, maybe four or five at the most, over her lifetime. As a result, the death of one orangutan mother has more impact on the survival of the species than would be the case if orangutan mothers bore litters once or twice a year. Nobody knows for sure, but in any case, numbers alone are not a measure of endangerment. Orangutans need forest; their survival as a wild species depends on the abundance and variety of mature fruit trees found only in the primary tropical rain forests of Borneo and northern Sumatra. Clearly, the main

illegally than are housed in all the world's zoos. Orangutans also are killed for what has become known as "bush meat," the consumption of primate flesh of all species not only by forest-dwelling indigenous peoples but by so-called gourmet diners in many Asian cities. As a result, orangutans were placed in appendix I of the Convention on International Trade in Endangered Species of Wild Fauna and Flora (CITES), which is the highest level of protection. In the orangutans' habitat countries, Indonesia and Malaysia, open trade of orangutans decreased markedly in the 1970s and 1980s. But the underground trade in living and dead animals has continued.

The orangutan species' problems began with timber. Thirty or forty years ago the forests of Kalimantan seemed to stretch forever. The low human population densities, vast forests, and traditional cultures ensured equilibrium. Traditionally, the native peoples of Borneo, the Dayaks, cleared small patches of primary forest by cutting and burning the trees to plant fields of dry rice. After some years the soil would become depleted, a new rice field was cleared, and the old one was abandoned. When natural secondary forest began to reclaim the abandoned field, the farmers planted vegetables and many species of fruit trees, some indistinguishable from their wild counterparts, as well as palm and bamboo. Researchers counted four hundred and twenty-five different plant species under cultivation in and around a single village in Java. The pattern in Dayak villages of Borneo was similar. These groves of planted trees, intermingled with secondary forest, mimicked the original rain forest, maintaining diversity. The human population was small, and the cutting, burning, and planting done to aid farming were small-scale activities. Until recent times, Borneo was a mosaic of vast stretches of primary tropical rain forest and small, scattered, cultivated groves in different stages of maturity. An opportunistic, albeit slow-moving species, the semisolitary orangutan was able to exploit these different forests, woods, and groves while avoiding much interaction with people. This has apparently been the case for a very long time, a fact confirmed by the relative scarcity of orangutan lore in local myth. It seems that people and orangutans stayed out of each other's way.

Like many indigenous peoples, the Dayaks are traditionally animists, who viewed the forest and especially the large trees as sacred. Out of reverence they left certain trees standing and certain areas of primary forest intact. In return, the forest provided wild animals for hunting and trees and plants for harvesting. The Dayak village of Pasir Panjang in Kalimantan Tengah, for example, was nestled amid mature wild fruit trees, vines bearing fruit and berries, trees with bee's nests, plants used for medicinal purposes, dyes, pesticides, and ropes or ties, as well as trees whose wood was prized for specialized purposes, such as making ax and machete handles, carvings, and containers. The trees and plants of the forest were owned communally and were subject to strong social controls that prevented overexploitation. Only plants that were known to produce new harvests were gathered. Among the Dayak of Pasir Panjang, for example, the legendary climbing skills of the young men allowed them to scale the tallest trees and gather ripe fruit high in the forest canopy; but custom demanded that at least one branch of ripe fruit must always be left behind for the tree.

An example of the change in atitude toward the primeval forest that has taken place in Indonesia can be found in the Mentawai Islands off the coast of western Sumatra. There, the primary forest remained essentially intact until the coming of

Christian missionaries and settlers from other islands who did not cherish the forest as did the original inhabitants. Local beliefs demanded that before a large tree could be felled a feast must be held, for which many boars had to be collected for slaughter. As in other rain forest cultures, pigs represented wealth, a valued resource to be spent only on special occasions. Unfortunately, once the local people saw missionaries and others cutting down tropical rain forest giants with impunity and suffering no ill consequences, they began dispensing with the rituals and the gathering of pigs and began felling trees at random. From a practical standpoint, they had to keep up. If they didn't cut down the trees, the newcomers would. Suddenly, the forest was up for grabs.

When I arrived in Kalimantan in 1971 I observed much the same scenario. Small local companies were beginning to harvest tropical hardwoods for commercial purposes. These small companies used slow hand-logging methods: men laboriously cut down hardwood trees with machetes and axes and then pulled the logs along simple, hastily constructed wooden tracks to the rivers for transport to the nearest town. If the logs were not used locally for construction, they were sold, unsawn and unmilled, for export. This practice, in retrospect, seems almost benign. But in the 1980s the situation changed. The government began awarding logging concessions to large lumber companies, the highest bidders. Well-funded consortiums brought in bulldozers, chain saws, and chipping machines. In terms of commercial exploitation, Indonesian Borneo, and Kalimantan Tengah, in particular, were latecomers, discovered when other tropical rain forests had been virtually logged to death. The profits from chopping up trees of every species, age, and size for wood products were much greater than the profits that had been made even from old, rare hardwoods that had been cut selectively. Clear-cutting became the rule. Meanwhile, local people lost the legal right to cut trees in their own forests. Contrary to what many outsiders believe, increased production did not mean increased opportunities for local people. Often, logging companies imported their own skilled crews as well as their superefficient machines.

Around the globe the exploitation of tropical rain forests for timber has increased exponentially. The orangutan is cursed by living in forests that are ideal for mass marketing to the industrialized world. Wood from tropical rain forests in other parts of the world cannot compete in the international timber market with the woods of Southeast Asia, whose qualities make them commercially ideal. The vast forests of the Amazon Basin include many species with dark, heavy wood that cannot be used for the plywood, peeled veneers, and light particleboard demanded by the world markets. Some African countries that used to boast vast tropical rain forests of their own are actually importing timber because their own forests either have disappeared or are inaccessible due to political and social chaos or lack of infrastructure.

Much to the detriment of orangutans and other native wildlife, the forests of Borneo (and, to a lesser extent, Sumatra) are a perfect source of wild wood. Trees that reach into the upper canopy consist of dozens of dipterocarp species, but these numerous species share common characteristics that allow them to be sold together as only a few major timber classes. A steady supply of wood organized into a few grades matches the needs of the buyers, who prefer to deal with only a few types of timber. In

The damaged earth peers from the forest, victim of massive gold panning on the upper reaches of the Sekonyer Kanan River.

the last two decades, Southeast Asia has become the world's leading source of tropical timber. Indonesia itself has become the world's largest plywood exporter. Far from being an economic backwater, Southeast Asia as a whole began exhibiting annual growth rates of 8 percent or 9 percent, which made this region an important emerging component of the fastest growing economic region in the world, the Pacific Rim. As was true in the more gradual industrialization of today's developed countries, the modernization of Southeast Asia has had a profound and almost entirely negative impact on endangered wildlife and their habitats.

The destruction and deterioration of tropical rain forest are not unique to Indonesia or Malaysia: it is a global problem. According to conservative estimates, at least half of 1 percent of the planet's remaining tropical rain forest vanishes each year. Some experts think that the destruction is several times greater because governments and conservation agencies do not count local, small-scale deforestation. If degradation of forest ecosystems is included, then at least 11 million hectares of forest are lost per year in the tropics. This translates into a loss of 20 hectares, nearly 50 acres, per minute.

Logging is not the only threat to the tropical rain forest. Agriculture is the main reason that these forest are cleared. In central Africa, for example, deforestation began with Westerners clearing plantations to produce rubber, bananas, and other cash crops for export during colonial and postcolonial times. Today's deforestation is the result of growing populations of local farmers cutting deeper and deeper into the forest to clear fields and grow crops to feed their families. In Central America, the Amazon Basin, and in the Atlantic coast forest of Brazil, mile upon mile of forest has been converted to pasture for cattle (some of which ends up in North America as fast-food hamburgers).

In Borneo, the introduction of commercial agriculture is relatively recent. Yet hundreds of thousands of hectares of forest have been cleared for palm oil plantations in the last few years alone. The oil palm is a native of Africa. When Jane Goodall began her studies at Gombe Stream, she noticed that the wild chimpanzees were attracted to the fruit of the wild oil palms that grew near her camp. Perhaps the palms had been planted by humans, but more likely they were part of the wild flora in the area. Oil palm does not grow wild in Kalimantan, but once planted it does well in the generally infertile sandy soils. Oil palm seedlings grow rapidly, beginning to produce after about three years. Now, large areas of forest are rapidly being cleared and burned to plant oil palm. Oil palm plantations produce greater annual income than selectively logging a tropical rain forest or even cutting it all down for timber.

Once the forest is cleared, the highly endangered orangutans have no place to go and no place to hide. They cannot simply vanish into thin air or disappear. Rather, attracted by the inner shoots of the growing oil palms, and later by the fruit itself, many become crop raiders. The result is predictable. When caught in a field, orangutans may be clubbed or stoned to death, for they are now regarded as a pest species. Plantation managers put a bounty on orangutan heads, but crop-raiding orangutans are not the only victims. In the mid-1990s, the going price for a pig tail was ten thousand rupiahs and for an orangutan head, fifty thousand rupiahs, at the time equal to a week or more of wages. If the victim is a mother, the orangutan bounty hunter can

LEFT: *Timber fuels the economies of Kalimantan's four provinces. Legal logging is a necessity, but determining what is legal and what is not has proved to be a nightmare for conservation officials of the region. Log yards like this are commonplace in the big towns of coastal Borneo.*

TOP: *A small boat slowly drags a raft of logs down the Sekonyer River to the Java Sea. Illegal logging operations near Tanjung Puting National Park damage the forest, but do not threaten the existence of the orangutans as much as clear-cutting, which totally destroys the forest.*

BELOW: *Pulling a ramin (*Gonistylus *species) log through burned forest, local workers extract a few logs each day. New, mechanized equipment is far more efficient and infintely devastating to the forest. A log such as this one may fetch as much as $30 or $40 when sold in town, but it will be worth hundreds by the time it reaches Japan or Western Europe.*

An Indonesian forest ranger stands guard over logs from an illegal operation inside Tanjung Puting National Park. Regrettably, the government is unable to increase the staff of forestry officials to more effectively patrol inroads into the park and other protected wilderness areas.

and no place to hide. They cannot simply vanish into thin air or disappear. Rather, attracted by the inner shoots of the growing oil palms, and later by the fruit itself, many become crop raiders. The result is predictable. When caught in a field, orangutans may be clubbed or stoned to death, for they are now regarded as a pest species. Plantation managers put a bounty on orangutan heads, but crop-raiding orangutans are not the only victims. In the mid-1990s, the going price for a pig tail was ten thousand rupiahs and for an orangutan head, fifty thousand rupiahs, at the time equal to a week or more of wages. If the victim is a mother, the orangutan bounty hunter can more than double his reward by capturing the infant or juvenile and selling the young orangutan on the black market. Most do not survive in captivity; some become bush meat. Judging by the ever-increasing number of infant and juvenile orangutans discovered along the roadsides and in local villages, many hundreds of orangutans have died, if not thousands. In the western part of one province in Kalimantan, an evil harvest of hundreds of infant orangutans per year is being reaped.

Tree plantations also are being established for timber and pulp and paper. Pepper is another important cash crop. Unlike palm oil and tree plantations, which are usually owned by large national and sometimes even international (but Asian) corporations, pepper plantations are more likely to be held by small local landowners and farmers. In terms of habitat destruction over large areas, pepper is less devastating than other types of plantations.

As large companies move in and local populations become involved in the cash economy, the primary forest simply becomes a nuisance. Logging and other roads are driven through to provide access to previously uninhabited and inaccessible places, and the silent majesty of the forest is violated. Increasing population pressure drives more people into the forest, both locals and newcomers, either as individuals or as part of government-funded development projects, and the traditional shifting horticulture gives way to large-scale forms of permanent agriculture. The balance tips; the equilibrium is lost.

Another threat to the stability of the forest is the lure of gold. In some areas of Borneo, especially the alluvial fans where great rivers meet in their flow toward the sea, the barren, sandy soil beneath the lush rain forest is flecked with gold. Indonesia has become one of the world's leading exporters of gold. And less than half of the twenty tons of gold exported each year comes from licensed mines. The bulk derives from wildcat, independent miners using pans, sieves, and diesel-powered water pumps and hoses to collect a few grams of gold on a lucky day. For the most part, these operations are small in scale, and the damage they do to the forest is somewhat limited. One exception is an area called Gedung Sintuk on the northwest border of Tanjung Puting National Park. There, gold fever has transformed the once pristine forest into a moonscape of deep gray pits shrouded by the hovering black fog of pump exhaust fumes. More than a thousand people have settled in the area in makeshift towns and villages that seemed to spring up overnight.

In addition to the clearing of trees for gold mining, further terrible damage is done by the use of heavy metals, especially mercury, to separate bits of gold from the sandy soil. These metals then run off in the wash water and contaminate the tons of

TOP: *Two women work a traditional gold site. One woman's face is covered with a rice and herbal paste that protects her from sunburn and smoothes the skin. The woman in the water tosses the dirt with her shovel onto the wooden tray where the earth is washed to separate the gold grains from the mud.*

BELOW: *The working conditions in gold-panning fields are harsh: men and women stand in foul water all day, burned by the broiling sun, beset by biting insects, and lacking clean water and hygiene. Competition for gold is so intense that if a hole contains evidence of the precious metal, people working it refuse to leave for days on end.*

TOP: *Modern gold-mining methods, with huge pumps, power hoses, and pneumatic drills for large crews, lay waste to the land far more quickly than in the past.*

BELOW: *An abandoned goldfield at the edge of a river looks like a moonscape.*

ter of a few yards the transparent waters of the tributary disappear into the cloudy murk of the polluted main river.

Logging, commercial plantations, and mining have caused at least half the forest of Borneo to disappear, to become terribly degraded, or to be replaced by secondary forest, scrub, or alang-alang grasslands that support far fewer species of wildife. One can peer into the future of Kalimantan by moving from Indonesian Borneo across the border into Malaysian Borneo, Sabah, and Sarawak. When I first visited Sabah in 1990, I was amazed by its beauty and astonished by its development—gas stations, direct dial telephones, faxes, and air conditioned shopping malls. At that time there were no telephone lines and only a few TV antennas in the small town in Kalimantan Tengah, where I lived when I was not at Camp Leakey. I found that an asphalt road ran right up to the gates of the forest reserve that sheltered the world's first orangutan rehabilitation program at Sepilok. Palm oil plantations covered the land. There was little space for wild nature outside the parks and reserves. At the conference I was attending, a local dignitary mentioned in his speech at the closing ceremony that the twinkling lights the people of Sabah looked up to at night were no longer the stars but the navigation lights of passing Japanese timber ships.

Since 1997 a terrible natural disaster has further diminished the forests of Borneo. During 1982 and 1983 what was then called the Great Fire of Borneo raged for almost a year. Up to that time, it was the largest forest fire recorded in human history. Four million hectares of lowland tropical rain forest were demolished, creating a thick blanket of smoky haze that closed airports as far away as mainland Asia. Sparked by the numerous small fires traditionally lit to clear rice fields, this great fire was considered brought on by El Niño, a climatic oscillation that warms the normally cool surface of the eastern Pacific Ocean. The rains that normally fall on Indonesia during the wet season instead fell into the waters of the equatorial Pacific, causing a prolonged drought.

In 1997 another long drought produced by El Niño caused a colossal forest fire that was fanned by relentless human activity. Massive deforestation, the result of legal and illegal logging on a massive scale and clear-cutting to make way for plantations, produced a disaster of epic proportions. The smoke was so dense and so widespread that people in Kuala Lumpur, Malaysia, and Bangkok, Thailand, felt its effects. President Suharto of Indonesia, in an unprecedented move, sincerely apologized for the fires to his own people and the people and nations of the region. While in 1983 the fires had been predominantly in Sabah and East Borneo, in 1997 they spread over the entire island of Borneo. Worst hit was the province of Kalimantan Tengah. Estimates of the damages are probably gross underestimates. This fire was perhaps more devastating even than the fire of 1983.

Apart from the blackened, smoldering landscapes and dense haze as far away as Darwin, Australia, one of the most visible consequences of the fire in Kalimantan Tengah was the flood of orangutans coming into captivity. Numbers more, especially mothers of infants and juveniles, were slaughtered. A foreign film crew investigating the fires found three captive orangutan infants in less than an hour in Palangka Raya, the provincial capital of Kalimantan Tengah. In one case, the badly malnourished

With lumbering and gold mining bringing increased affluence to Kalimantan, riverside communities like this one are growing beyond their old borders, infringing even further on forest habitats and placing great strains on natural resources.

infant's cage was partially covered by the drying pelt of his mother. The illiterate farmer proudly demonstrated his technique for killing orangutans fleeing on the ground from the decimated forest. At least these farmers were eating the orangutans they slaughtered. The film crew was so shaken that the members stopped inquiring about orangutans. The following week another group discovered seven newly captured infants in a small town in the province. All ten infants were confiscated, and plans were made to send them to a rehabilitation program.

Rehabilitation, that is, the planned return of captive orangutans to the wild, has been a controversial idea. Scientists, government officials, and conservationists, as well as many people who have never seen an orangutan in the wild, strenuously argue the pros and cons of rehabilitation and the merits of various programs. This preoccupation with rehabilitation is a symptom of the failure to get to the underlying causes of the orangutan crisis. Rather than endlessly debating the detailed aspects of rehabilitation programs, we should be talking about how to stop the flow of captive orangutans and the destruction of orangutan habitat. Evaluating the scientific validity of rehabilitative procedures is important, but the evaluation depends on the goals of rehabilitation.

In 1971, my former husband, Rod Brindamour, and I established a rehabilitation program at Tanjung Puting Reserve (now a national park). It was the first program in Kalimantan and it paralleled a similar program established at the same time in northern Sumatra at Ketambe in Gunung Leuser Reserve (now also a national park). Our reasons for establishing the rehabilitation program at Camp Leakey included: protecting the integrity of Tanjung Puting Reserve by highlighting the rehabilitation program; removing orangutans from the commercial or pet trade; improving the quality of life for wild-born, ex-captive orangutans; educating local people and officials, as well as foreign visitors; and establishing tourism appeal as a means to aid the ultimate cause. Most important, we sought to preserve the forest around Camp Leakey and to protect the wild orangutans living or released there. In this, we succeeded. The northern part of the park, in the vicinity of the Sekonyer Kanan River, has been safeguarded and is considered by many the safest refuge for wild orangutans in Kalimantan.

The program associated with Tanjung Puting National Park has released more than two hundred orangutans to the wild. Successful orangutan rehabilitants engage in species-appropriate behavior, reproduce successfully, and, in the case of females, rear offspring. One of these youngsters, Siswi, has herself became a mother. At Camp Leakey, ex-captive orangutans and their wild offspring occasionally return for the daily feedings, especially when fruit in the forest is scarce. Wild orangutans also appear at the feedings from time to time, but this is rare. Some ex-captives also return for stimulation, for social interaction with other orangutans, and even humans. Not a single orangutan who has been released has stayed in our camp totally dependent on daily feedings, much less human companionship.

When ex-captive male orangutans approach maturity, they invariably leave the area. Some return unpredictably after long absences, but most do not—or, if they do, they return as unrecognizable cheekpadders. Now and then I think I recognize a mature male who disappeared as a juvenile or subadult, but in many cases I can't be sure. Female ex-captives are somewhat more likely to remain in the vicinity and to appear at feedings with a new infant draped on their heads or a juvenile in tow. Some female ex-captives appear periodically for a few years, then vanish. A few are regular visitors, who seem to have made the area around Camp Leakey their home base. In general, wild orangutan females, such as Priscilla, whom I have followed and studied for more than twenty-five years, stay around. The only behavior that distinguishes mature wild orangutans from mature ex-captive orangutans, whose natural lives were interrupted by separation from their mothers and varying terms in captivity, is that some are less wary of human beings. But they are the exceptions. Of the two hundred ex-captive orangutans we have released at Camp Leakey and at other sites in the area, perhaps a dozen boldly approach human visitors (usually looking for food) or behave as if the visitors were not there. The great majority go wild, in the sense that they put their experiences with humans behind them.

The survival rate in our program has been encouraging. In a tamarin rehabilitation program in South America, a 50 percent survival rate for the released animals was considered a success. At Tanjung Puting the known survival rate is much higher. An exact figure is difficult to calculate because, like wild orangutans, the ex-captives come and go and, even when nearby, often elude identification by human observers. A good example is Bagong, who arrived at Camp Leakey as a grown male who had been held in a zoo in Java. He had a very distinctive face. On release, he immediately disappeared into the forest, then suddenly reappeared at Camp Leakey one and a half years later and visited (very infrequently) thereafter. Almost twenty years later he still occasionally appears at Camp Leakey, clearly older and battered but still alive.

Throughout Borneo and Sumatra over the last thirty years rehabilitation programs have released at least 800 wild born ex-captive orangutans: more than 200 at Bohorok, 200 at Tanjung Puting, 200 at Sepilok, and 200 in the now closed program at Kutai and through Wanariset in East Borneo. Not surprisingly, a few of these released individuals have caused problems for humans, such as raiding buildings for food or biting people. It is remarkable, however, that so many wild-born, ex-captive orangutans have adapted to life in the wild. When Joy Adamson and her husband returned a single lion, Elsa, to the wild in the 1960s, conservationists and animal lovers were awed. The 1993 film *Free Willy*, a fictional account of the return of a single killer whale (played by a killer whale) to its natural habitat, made audiences cheer. Multiply the story of Elsa or Willy eight hundred times.

Critics have argued that such programs as ours may have negative impacts on wild orangutan populations: that the ex-captives may spread disease, compete for food resources, or otherwise affect the wild orangutan's ranging patterns. Hypothetically, all these things are possible. In reality, however, there is no evidence that any of these potential hazards have occurred. Have wild orangutans been contaminated by diseases from ex-captives? No. Have their normal behavior patterns changed because of the introduction of ex-captives? Again, and based on my ongoing research, the answer is no. Has anyone produced evidence that malnutrition and death rates are higher in areas where ex-captives have been introduced? Yet again, the answer is no. The death rates for wild orangutans are highest when their forest habitat is demolished. The practical reality is that the presence of orangutan rehabilitation programs helps protect the forest for the wild orangutans.

Multiply the story of Elsa or Willy eight hundred times.

Critics have argued that such programs as ours may have negative impacts on wild orangutan populations: that the ex-captives may spread disease, compete for food resources, or otherwise affect the wild orangutan's ranging patterns. Hypothetically, all these things are possible. In reality, however, there is no evidence that any of these potential hazards have occurred. Have wild orangutans been contaminated by diseases from ex-captives? No. Have their normal behavior patterns changed because of the introduction of ex-captives? Again, and based on my ongoing research, the answer is no. Has anyone produced evidence that malnutrition and death rates are higher in areas where ex-captives have been introduced? Yet again, the answer is no. The death rates for wild orangutans are highest when their forest habitat is demolished. The practical reality is that the presence of orangutan rehabilitation programs helps protect the forest for the wild orangutans.

One of the ways that this is happening is through ecotourism (also known as value or nature tourism). Ecotourism has been promoted as at least a partial solution to the problems faced by endangered species. It injects a positive economic element into the equation of conservation. Endangered species earn their keep by attracting visitors. Local people are employed to maintain the sites, to work in facilities established to feed and house the visitors, and to guide and guard travelers. A good case is Sepilok, where in the early 1990s, seventy thousand people visited the site. In Tanjung Puting the growth of tourism was pioneered by Earth Watch, a Boston-based group that found volunteers willing to pay their own way to come to help Camp Leakey's programs, and by Orangutan Foundation International's study groups. This generated employment for local people and helped create institutional and governmental, as well as local, support and commitment for orangutan conservation and protection of the forest. The ecotourists also became goodwill ambassadors for the work being done at Tanjung Puting and, by extension, elsewhere.

Critics of rehabilitation programs ignore the fact that in many ways orangutans are ideal rehabilitants. Because they are semisolitary in the wild, orangutans do not have to be accepted into a family, troop, or community, as gorillas, monkeys, or chimpanzees would have to be. As a result, the introduction of ex-captive orangutans to the forest is more successful: the rehabilitants manage perfectly well on their own in the wild, and their added presence does not disrupt the normal social patterns of wild orangutans. Equally significant, the low birthrate of orangutans means that local wild populations are not suddenly swamped by litters of young, hungry animals.

Under normal conditions, orangutans do not interfere with human activities. To the contrary, as evidenced by the difficulty Western scientists had even locating wild orangutans for study, orangutans avoid people, exploiting the riches of a forest canopy that without technology is beyond our reach. What critics seem not to recognize is that returning orangutans to the forest through rehabilitation programs save lives, the lives of the ex-captives themselves and even the lives of the wild orangutans because of the newly safeguarded habitat.

A slash and burn cultivator's hut sits by a river bank on the edge of a field cleared from the rain forest. Burning the dead trees provides ash as fertilizer for the dry rice crop. As available land diminshes, the traditional shifting cultivation of former years, based on slash-and-burn methods, is gradually changing to a permanent agriculture that cannot be sustained long term.

Wild orangutan populations are under siege, but the villain in this real-life tragedy is not the rehabilitation programs and the introduction into the forests of ex-captive orangutans but deforestation. In Sabah, starving wild orangutans have been encountered at the edge of rubber and other plantations. In Kalimantan Tengah desperate wild orangutans are found (and often slaughtered) within palm oil plantations miles from the nearest wild tree. Like homeless people, homeless orangutans are a nuisance, an eyesore, an uncomfortable tug on the conscience. Society prefers either to turn its back on this population or to eliminate it, whichever provides the easiest solution.

The simple solution—saving the tropical rain forests where orangutans make their homes—is the most difficult to implement. There is no doubt that the orangutan stands at the edge of extinction. A special report issued by the Environmental Investigation Agency in July 1998 estimated that the wild population of orangutans has declined by 50 percent in the last decade. Somewhere between fifteen thousand and twenty-five thousand animals are believed to survive in the wild, and the large majority of those are living in Borneo and Sumatra. Up to 80 percent of this forest habitat has been lost over the past twenty years. As long as orangutan habitat is being demolished, orangutans will either be killed or will enter captivity, a fate akin to death. In Kalimantan they do not survive well in captivity. Putting them back into the forest through rehabilitation programs saves lives. If some tropical rain forests are not

In a modern Dayak long house a woman holds a juvenile male orangutan who, since his infancy, has been her substitute child. Against the wall leans a traditional hunting weapon, the blowgun, which, with its poison-tipped darts, may have been the weapon that killed this orangutan's mother. In the past Dayaks ate orangutan meat, and many continue to do so today.

LEFT: A Balinese shop keeper smiles as he holds up for sale an engraved orangutan skull carved in a traditional Dayak motif. Carved orangutan skulls in the mid-1990s sold for anywhere from several hundred to several thousand U.S. dollars despite the fact that their sale was illegal.

set aside from development, and if orangutans are not protected from exploitation, they will become extinct in the wild. This extinction will not be as complete as, for example, that of the North American passenger pigeons, but like the European bison. In that case, a few herds of semidomesticated bison survive in forest lands on the Polish–Lithuania border, but as an ecological force, the European bison vanished hundreds of years ago. There are no more wild European bison.

To save the orangutan, the forest must be preserved—an improbable although not an impossible goal. As a species, we have thrived by achieving what seems to be impossible. Yet here, the possible continues to elude us. In the end the solution to orangutan extinction will be piecemeal: a patchwork of economic, political, cultural, and social negotiations and compromises. Someday soon there will be no more orangutans except in national parks and reserves. One could say that the human species had outcompeted one of its nearest hominoid relatives. (And will likely do so with the others that remain.) Won't we become lonely on a planet endlessly spinning, contemplating all that once was and all that we once shared with the wild orangutan? We need to save the forest. In so doing, we save the orangutan.

A Borneo family keeps a juvenile orangutan chained in the yard. Government efforts to confiscate such inappropriate pets have been only partly successful.

At a convenience shop in Taiwan a pet orangutan amuses customers. Taiwan has been a leading destination for captured orangutan infants and juveniles.

The trade in captured orangutans continues despite worldwide efforts to block it. Seeing captive animals such as these gives vivid evidence of the cruelty of removing such creatures from their native habitats.

For nearly thirty years, the author, who is seldom seen at Camp Leakey without an infant orangutan clinging to her, has studies wild adult orangutans and rescued orphan youngsters.

OPPOSITE: *A rescued orangutan infant from Orangutan Foundation International is moved by a volunteer deeper into the forest to further his burgeoning skills in the wild. Here, near a simple camp in the primary rain forest, he will practice climbing, nest-building, and foraging for food. Volunteers are encouraged to wear masks and to wash their hands frequently in order to protect the orangutans from becoming infected by human diseases.*

PENN

Biruté Galdikas (background) guides a volunteer carrying an ex-captive young orangutan on a wet walk into the forest to the youngter's new home.

OPPOSITE: *Deep in the forest, at a simple camp, volunteers take a needed break from their work with ex-captive orangutans. The Orangutan Research and Conservation Program, based at Camp Leakey as well as in nearby areas, and directed by Biruté Galdikas, has successfully returned more than two hundred orangutans to the wild.*

At Camp Leakey, four orangutan juveniles are fed bananas and milk. Since 1992 the Forestry Department has taken over the public feedings of the wild-born, ex-captive orangutans in the Tanjung Puting National Park. The day laborer employed by the government holds a stick and slingshot to drive off older ex-captive orangutans who may be as hungry as the young juveniles. This practice has sometimes elicited criticism from visitors.

A group of visitors to Tanjung Puting National Park enjoys a chance meeting with an orangutan mother and child, who have likely come out of curiosity to see what is happening

Volunteers from the United States and elsewhere make a considerable contribution to the work at Camp Leakey and other sites. Staying for several weeks at a time, these unpaid workers, among them Tiffany Boswell from Florida (opposite), pitch in at any number of necessary chores, taking the burden off the professional staff. Over the years, hundreds of such volunteers, some of them coming on paid "vacations" as eco-tourists through Earth Watch and such organizations, have made their way to Camp Leakey. Still others have come as volunteers through Orangutan Foundation International and made invaluable contributions of tens of thousands of work hours. Seen here also is photographer Karl Ammann, who has made an enormous contribution to this book by providing his photographs. Ammann, who lives in Kenya, has been to Camp Leakey and elsewhere in Borneo numerous times and is a leading advocate for wildlife conservation, with a particular concern for the great apes. He conducts a worldwide campaign to make more widely known the terrible practice of killing apes for restaurant and local food, otherwise known as "bush meat."

Nancy Briggs plays with an orangutan at a small sanctuary established outside Indonesia by a former Camp Leakey volunteer.

TAMAN NASIONAL TANJUNG PUTING
SEBAGAI KAWASAN KONSERVASI
YANG DILINDUNGI OLEH UNDANG-UNDANG
/PERATURAN PEMERINTAH, DALAM KA -
WASAN INI DILARANG :
- MASUK TANPA IJIN
- MERUSAK HUTAN/MENEBANG POHON
- MEMUNGUT HASIL HUTAN (ROTAN, JELUTUNG DSB)
- MELAKUKAN PENAMBANGAN/GALIAN TANAH
- BERBURU BINATANG/MENANGKAP IKAN

The innocent beauty of orangutans in the wild should provide evidence enough of their value to humankind. That they are near relatives of humans is far from the only reason for their protection.

ACKNOWLEDGMENTS

All of the royalties generated by the sale of this book go directly to the Orangutan Foundation International to help save orangutans and the forest. (OFI is located at 822 South Wellesley, Los Angeles, California 90049. (310) 207–1655; fax: (310) 207–1556; e-mail: redape@ns.Net; website: http://ofi.Net)

First we wish to thank the government of Indonesia and its officials for their continued support of the Orangutan Foundation International's mission: the late Minister Soesilo Soedarman, Minister Ali Alatas, former Minister Joop Avé, former Minister Djamaludin Suryohadikusomo, former Minister Ir. Sarwono Kusumaatmadja, former Ambassador Abdul Rachman Ramly, former Ambassador Dr. Arifin Siregar, former Minister Dr. Soedjarwo, Prof. Emil Salim, Mr. Sunten Manurung, Ambassador Dr. Doro Djatun Kuntjoro-Jakti.

For patience, understanding, and support, the authors wish to thank their immediate families: We thank the family of Dr. Biruté M. F. Galdikas: Pak Bohap Bin Jalan, Binti Brindamour, Fred Galdikas, Jane Galdikas, Antanas Galdikas, Filomena Galdikas, and Aldona Galdikas.

We thank the family of Dr. Nancy Erickson Briggs: Rod Briggs, Eric Briggs, Nicole Briggs, Otto Erickson, Jan Steadman, and Sonja Staley.

We cannot thank Karl Ammann enough. We believe Karl is a photographer of great genius and generosity. He cares for orangutans and their survival, and his contribution of extraordinary photographs for this book is greatly appreciated.

Special thanks go to contributors, including Dr. Gary Shapiro, Mrs. Inggriani Shapiro, Ann Levine, Noel Rowe, Michael L. Charters, Tiffany Boswell, Rod Brindamour, Isabella Rossellini, and Julia Roberts. We also sincerely thank Ann Levine for her editing help and advice. Her love of orangutans is exemplary.

Our sincere gratitude also is extended to OFI board members: John Beal, Tiffany Boswell, Caroline Gabel, Steven Karbank, Norman Lear, Ashley Leiman, Eric Raymond, Gary Shapiro, Barbara Spencer, and Gerald and Pearl Sugarman. We thank close OFI friends such as Kevin and Linda Nealon, Ed Begley, Stefanie Powers, Betty White, Dr. Jane Goodall, Gordon Getty, Dr. Lillian Rachlin, Blanche Whittey, Robert Wilkie, Dr. Sumitro Djojohadikusumo, Dr. Karin Lind, Charlotte Grimm and Pak Uil Otol, Michelle Dujmovic, Andrea Gorzitze and Ralph Arbus.

Our gratitude also is extended to these supporters of this book: Mimi Abers, Glenda Adams, Kay Bassford, George S. Bell, Robert J. Benke, Dottie Berke, Gary Blond, William H. Breitmeyer, Jamie Brown, Patsy Cashmore, Jan Davies, Helen Davis, Jim De Lara, Jack and Karen Derrico, William Disher, Suzy Dorr, Mandy Dunford, Claire Easby,

Debra Gamble, Carol Gee, Louise Geist, Laura Gerwitz, Cindy Gibat, Ellen Goff, Elizabeth Groat, Ralph R. Gut, Jackie Hanrahan, Lou Harrell, Valerie M. Hart, Al and Marka Hibbs, Kathy Irwin, Kerry Jess, David Lappen, Rebecca Levine, Jan V. Levitan, Don and Margie Mennell, Audrey Mertz, Claudia Olesniczak, Carl Palazzolo, Liselotte Paulson, Carol S. Piligian, Silvia Reiche, Marty Roper, Harry L. Rossi, Noel Rowe, Miriam F. Shapiro, Barbara Shaw, D. P. Spencer, Graeme Strike, Maureen Taubman, Karen Taubner, Jo Anne Tilzhman, Mirian L. Trogdon, Dr. Maylene Wong, Kimberly Wood, Sharon Yanish, and Dr. Anne Zeller.

We also wish to thank former Universitas Nasional students: Suharto Djojosudarmo, Jaumat Dulhaja, Endang Soekara, Barita Manulang, Dr. Yatna Supriatna, Zackie Ichlas, and Edy Hendras.

In addition, we sincerely acknowledge support from the following:

Paula Adams, Scott Atkinson, Allen Altcheck, Winifred Barrows, Jonathan Boswell, John Boswell, Julie Boswell, Earl Holliman, Ben Benniardi, Corazon Bryan, Ed Rosenblum, Paul and Judy Clark, Jon Coe, Martine Collette, Kathie Comerford, Lisa Couturier, John Paul De Joria, Dr. Bill Doyle, Thaya DuBois, Larry Ellison, Neva Folk, Tom Gause, Evelyn Gallardo and David Root, Wendy and Russ Gilbertson, Dr. Richard Glassberg, Drs. Ken and Jan Gordon, Bonnie Hall, Junann Holmes, Hon. Steven Horn, Cathryn Hilker, Tippi Hedren, Wendy Hoole, Tim and Cindy Hunter, Laura Hyman, Jennifer Ingle, the late Lorraine Jenkins, Alistair and Heather Kent, David Koch, David Kleeman, Sari Jayne Koshetz, Gladys Lacy, Elaine and Ken Leventhal, Hon. Andrew and Nicolette Levin, Joan Lissaur, Edward and Aileen Master, John MacKinnon, Steve Mills, Manual Mollinedo, Dr. Mark Morris, Cam Muir, Judge Kenton Musgrave, Ron Orenstein, Rosalie Gann, Jindra Ottenfeld, Rosemary Olszewski, Hank Palmieri, Carmen Paredes, Nancy Pearlman, John Pearson, Claire Pollack, Sandy and Harold Price, Patti Ragan, Bill Raffin, Rosemary Raitt, Susan Raisin, Ken Redman, Bill Richardson, April Riddle, Trudy Rideout, Nick and Nancy Rutgers, Dr. Ed Sleeper, Mary Smith, Mark Starowicz, Sally Spencer, Byron Stafford, Imo Sundin, Siegfried and Roy, Maurine Taubman, Ann Thompkins, John and Susan Thorton, Linh Tran, Cynthia Tsai, Sherryl Valpone, Liz Varnhagen, Jim Watt, Neal Weisman, the late Leighton Wilkie, Larry Williams, Gretchen Wyler and Betsy Yap. And from Dean Dorothy Abrahamse, Jack Blaney, Leif Cocks, Nora Fraser, Rosalie Gann, Suwanna Gauntlet, Matt Groenig, David Hasselhoff, Phil Hobler, Peter Knights, Peter Max, Robert Maxson, Shirey McGreat, Alex Pacheco, Glenn Shiigi, Richard Shutler, Julie Tulang, Linda Wallace-Grey, Richard and Gloria Wurman, Mayor Stephen Yamashiro. We also thank literary agent Mike Hamilburg for his generous efforts on our behalf.

Finally, our deepest gratitude goes to Robert Morton, who had the vision to understand what this book could mean to the world and to future generations as the orangutan disappears from the wild.

Index

Page numbers in *italics* indicate photographs.